Document-Image Related Visual Sensors and Machine Learning Techniques

Document-Image Related Visual Sensors and Machine Learning Techniques

Editors

Kyandoghere Kyamakya
Fadi Al-Machot
Ahmad Haj Mosa
Jean Chamberlain Chedjou

MDPI • Basel • Beijing • Wuhan • Barcelona • Belgrade • Manchester • Tokyo • Cluj • Tianjin

MDPI

Editors
Kyandoghere Kyamakya
University Klagenfurt
Austria

Fadi Al-Machot
Alpen-Adria-Universität
Klagenfurt
Austria

Ahmad Haj Mosa
Alpen-Adria-Universität
Klagenfurt
Austria

Jean Chamberlain Chedjou
Alpen-Adria-Universität
Klagenfurt
Austria

Editorial Office
MDPI
St. Alban-Anlage 66
4052 Basel, Switzerland

This is a reprint of articles from the Special Issue published online in the open access journal *Sensors* (ISSN 1424-8220) (available at: https://www.mdpi.com/journal/sensors/special_issues/Document-Image_Visual_Sensors).

For citation purposes, cite each article independently as indicated on the article page online and as indicated below:

LastName, A.A.; LastName, B.B.; LastName, C.C. Article Title. *Journal Name* **Year**, *Volume Number*, Page Range.

ISBN 978-3-0365-3026-0 (Hbk)
ISBN 978-3-0365-3027-7 (PDF)

Contents

About the Editors

Kyandoghere Kyamakya

Kyandoghere Kyamakya is currently a full professor of transportation informatics and deputy director of the Institute for Smart Systems Technologies at Universitaet Klagenfurt in Austria. He is actively conducting research involving modeling, simulation, and test-bed evaluations for a series of concepts applied amongst others, but not exclusively, in the context of Intelligent Transportation Systems. In his research does involve a series of fields such as nonlinear dynamics, systems science, machine learning / deep learning, nonlinear image processing, neurocomputing and partly telecommunications systems. He has co-edited more than 8 books and has so far published more than 100 journal papers and some hundreds conference papers.

Fadi Al-Machot

Fadi Al-Machot finished his PhD in computer science at Alpen-Adria University Klagenfurt in November 2013 and his habilitation in applied computer science at the University of Lübeck in 2020. As a researcher, he developed different algorithms and approaches in the areas of complex event detection in multimodal sensor networks, advanced driver assistance systems, human cognitive reasoning, and human activity and emotion recognition. His work is patented and published in different international conferences and Journals. He is currently a senior data scientist at Leibniz Lung Center – Research Center Borstel.

Ahmad Haj Mosa

Ahmad Haj Mosa developer in the team of Digital Services at PwC Austria. He is also a researcher and a lecturer at the Institute for Smart System Technology (IST) at the University of Klagenfurt, Austria. His research area focus lies on Augmented Intelligence and Explainable Deep Learning, and Self-Driving Cars. And his research interests include machine vision, machine learning, applied mathematics, and neurocomputing. He has developed a variety of methods in the scope of human-machine interaction and pattern recognition.

Jean Chamberlain Chedjou

Jean Chamberlain Chedjou is currently an Associate Professor at the Institute for Smart Systems Technologies, Universitaet Klagenfurt, Austria. He is conducting research in the field of dynamic systems in traffic engineering. His current research interests include nonlinear dynamics in intelligent transportation systems (ITS), applications of neural networks and cellular neural networks in ITS, electronics circuits engineering, and graph theory. He has been serving as a Reviewer in several journals, including IEEE ACCESS, the IEEE TRANSACTIONS ON NEURAL NETWORKS AND LEARNING SYSTEMS, the IEEE TRANSACTIONS ON CIRCUITS AND SYSTEMS, the IEEE TRANSACTIONS ON COMMUNICATIONS, NEURO-COMPUTING, NONLINEAR DYNAMICS, SENSORS, the International Journal of Bifurcation and Chaos, the Journal of Applied Physics, the AEU - International Journal of Electronics and Communications

sensors

…rial

…cument-Image Related Visual Sensors and Machine …arning Techniques

…ndoghere Kyamakya [1,*], Ahmad Haj Mosa [1], Fadi Al Machot [2] and Jean Chamberlain Chedjou [2]

1 Institute for Smart Systems Technologies, Faculty of Technical Sciences Universitaet Klagenfurt, A9020 Klagenfurt, Austria; ahmad.haj.mosa@pwc.com
2 Research Center Borstel—Leibniz Lung Center, 23845 Borstel, Germany; falmachot@fz-Borstel.de (F.A.M.); Jean.Chedjou@aau.at (J.C.C.)
* Correspondence: kyandoghere.kyamakya@aau.at

Document imaging/scanning approaches are essential techniques for digitalizing documents in various real-world contexts, e.g., libraries, office communication, management of workflows, and electronic archiving. Such a digitalization step plays an important role in decreasing costs and increasing the efficiency of document management systems.

Document management systems require document imaging/scanning approaches to convert hard-copy documents/images into digital files. However, document management systems are complex systems consisting of database servers and any document analysis related processes. The term document management refers to the database-supported management of electronic documents. A basic application of document management in the narrower sense is the digital files, in which information from various sources is either extracted or fused and refers to multiple system categories and their interaction in the broader sense.

Furthermore, the added value of such systems arises when documents have to be retrieved and/or analysed after some time due to legal requirements, and such a retrieval/analysis can be avoided or be related to financial penalties that can be significant for the industry. Moreover, costs and efforts can be reduced by retrieving documents. Increasingly, document imaging systems are being used as the base for organizational programs. The completion of tasks, orders, etc., is thus supported in logical and temporal sequences as workflows.

Since the conversion is not merely an image, Optical Character Recognition (OCR) is consecutively involved in recognizing and extracting the information contained in the document images. The documents can then be indexed and the extracted information can be transferred to a document management system for further processing. However, the OCR system does not show promising performance whenever images might be curved, distorted (e.g., by noise, blur, low contrast, and shadow), skewed, or have insufficient resolution, resulting in the loss of valuable image assets for character identification. Particularly hard distortion conditions occur nowadays when document images are acquired by using smart phone cameras. This means that while the image is accessible, the document might, however, not always be clearly readable.

In the state-of-the-art, there are many approaches to overcome the challenges of digital imaging/scanning systems: for example, utilizing self-learning systems with similarity/embedding vectors, neural models, and deep learning. Furthermore, pattern recognition can be used in two ways: (a) to determine the location of a predefined pattern in a larger image area, e.g., in a pick-and-place application where a vision system finds the object or the bar code and transmits the position to a robot; and (b) to focus classification on the nature of the visible object at a given location, e.g., in the case of text recognition where the position of each character is known but where it is necessary to determine which letter or digit is present.

Generally, high quality captured document images are required due to a series of challenges related to the performance of the visual sensors and, for camera-based captures,

…tion: Kyamakya, K.; Haj Mosa, …; Machot, F.A.; Chedjou, J.C. …ument-Image Related Visual …sors and Machine Learning …niques. *Sensors* **2021**, *21*, 5849. …s://doi.org/10.3390/s21175849

…eived: 17 June 2021
…epted: 17 August 2021
…lished: 30 August 2021

difficult external environmental conditions encountered during the sensing (image ca ing) process. Such document images are mostly hard to read, have low contrast, and corrupted by various artifacts such as noise, blur, shadows, spot lights, etc., just to na few. To ensure an acceptable quality of the final document-images that can be perfe digitalized and involved in various high-level applications based on digital docume the sensing process must be made much more robust than the raw capture result gener by a purely physical visual sensor. Thus, the physical sensors must be virtually augme by a series of additional pre-processing and/or post-processing functional blocks, w mostly involve, amongst others, advanced machine learning techniques.

This book emerging from the Special Issue "Document-Image Related Visual Sens and Machine Learning Techniques" can be viewed as a result of the crucial need document management systems. Such technologies are being applied in various field different domains and parts of the world to address challenges that could not be addres without the advances made in these technologies. The Special Issue includes nine pa submitted in response to the call for papers. The Special Issue includes impactful pa that present scientific concepts, frameworks, architectures and innovative ideas on sens technologies and machine-learning techniques to overcome the challenges of docum imaging/scanning, test detection, text recognition and documents clustering.

Overall, these papers can be grouped into the following three categories/groups:

- Visual Sensing;
- Document scanning and imaging;
- Document clustering and classification.

1. Visual Sensing

In [1], the authors propose a sensing concept for reliably classifying different ty of houses. For this challenging endeavour, they propose/introduce a novel convolutio neural network architecture involving multi-channel features extraction. The develop deep-learning model was trained with 600 images, verified with 200 images, and test with 400 other images. The performance (accuracy, precision, and so on) reached by proposed CNN model is at least 8% higher than that of the related models from the previo state-of-the-art, which have been involved in the rigorous benchmarking.

The authors of [2] suggest a composite filtering system for using consumer dep cameras at close range. The proposed method comprises three key components whi work together to remove various forms of noise. The system is GPU-accelerated and do not use window smoothing. The proposed approach has been tested by using both Kin v2 and SR300. The results demonstrate promising results and have exceptionally hi real-time accuracy, allowing it to be used as a pre-processor for real-time human-comput interaction and real-time 3D reconstruction.

2. Document Scanning and Imaging

Given the wide range of image binarization methods available and their vario implementations and image types, it is not easy to consider a single standardized thresho approach to be the right option for all images. There is still a lack w.r.t. deciding whi binarization methods are prone to increase OCR accuracy. As a result, the concept using robust combined steps is discussed in the work presented in [3] , whereby th benefits of different techniques are integrated/merged though including some recent suggested approaches focusing on entropy filtering and a multi-layered stack of regio The experimental results obtained for the WEZUT OCR Dataset, a dataset of 176 no uniformly illuminated text images, clearly confirm both the feasibility and utility of th proposed solution, resulting in substantial improvement in recognition accuracy.

The work in [4] proposes a low-cost scanner for capturing multispectral paper image Here, the authors modify a sheet-feed scanner by adding an external multispectral lig source made up of narrow-band light-emitting diodes to its internal light source (LED The modification does show promising results, coupled with compactness and low cos

The prototype design can be transformed into a fully functional portable product that can be used for multipurpose document analysis.

3. Document Clustering and Classification

In [5], the authors propose a scene text recognition algorithm using a text location correction (TPC) module and an encoder-decoder network (EDN) module. The TPC module converts the slanted text unto a horizontal text, and the EDN module then identifies the content of the flat text. For evaluation, the authors used both the ICDAR2013 and IIIT5K datasets. The experiments and the related evaluation results show promising results, and they additionally show that the proposed approach is capable of recognizing a wide range of odd text. The proposed two network modules improve the precision of abnormal scene text detection according to ablation studies.

The paper [6] introduces a Deep Convolutional Neural Network (DCNN)-based real-time supervised learning strategy for document classification that aims to reduce the influence of negative document image issues such as signatures, labels, logos, and handwritten notices. The authors propose a data augmentation strategy that uses the secondary dataset RVL-CDIP to normalize the imbalanced dataset. DCNN features are extracted using the VGG19 and AlexNet networks that are then fused, optimized, and modified by removing the redundant features using the Pearson correlation coefficient-based technique. The proposed approach is evaluated on the Tobacco dataset, whereby it shows promising classification results using a cubic support vector machine classifier.

In [7], the authors propose a text recognition Convolutional Neural Network (CNN) architecture that is adaptive to text scale to solve this problem. They use multi-stage convolution layers to extract multi-resolution feature maps in order to avoid missing details and to keep the feature size constant. The evaluation of the proposed model is performed using 7152 natural scene images containing texts. The main improvement is to introduce a multiple Region Proposal Network (RPN) to detect texts from different resolution feature maps. The suggested system outperforms the faster R-CNN by more than seven points on the F-score in the conducted experiments. Furthermore, the proposed approach produces findings that are similar to those of other methods. As a result, they have comprehensively tested the efficacy of the proposed approach, especially for text scales.

In [8], the work proposes a clustering approach in Wireless Multimedia Sensor Networks (WMSN). The aim is to overcome the problem of feature extraction from incomplete data. Therefore, the researchers of this work suggest (a) the use of the optimally constructed variational autoencoder networks for feature extraction from incomplete data, (b) improving the clustering output by using the High-Order Fuzzy C-Means algorithm (HOFCM), and (c) recovering the missing data by using low-dimensional latent space of the variational autoencoder. The experiments on different datasets show that the proposed algorithm improves the clustering accuracy for incomplete data and fills in missing features properly.

The research in [9] contributes in detecting and recognizing charts. The proposed system automates the process by using perspective detection and correction. These methods transform a blurred and noisy input into a simple chart that is ready for data extraction. Different models have been tested for classification and detection, e.g., Xception, ResNet152, VGG19, MobileNet, RetinaNet, and Faster Region-Based Convolutional Neural Network (R-CNN). The authors collected 21,099 chart images from Google, Baidu, Yahoo, Bing, AOL, and Sogou for evaluation. The total number of charts' classes is 13. The obtained results and the evaluation metrics in this work show that chart recognition methods can be applied for real-world applications.

Conflicts of Interest: The authors declare no conflict of interest.

References

1. Tavakkoli, V.; Mohsenzadegan, K.; Kyamakya, K. A Visual Sensing Concept for Robustly Classifying House Types throu Convolutional Neural Network Architecture Involving a Multi-Channel Features Extraction. *Sensors* **2020**, *20*, 5672. [Cross [PubMed]
2. Dai, Y.; Fu, Y.; Li, B.; Zhang, X.; Yu, T.; Wang, W. A New Filtering System for Using a Consumer Depth Camera at Close Ra *Sensors* **2019**, *19*, 3460. [CrossRef] [PubMed]
3. Michalak, H.; Okarma, K. Robust Combined Binarization Method of Non-Uniformly Illuminated Document Images Alphanumerical Character Recognition. *Sensors* **2020**, *20*, 2914. [CrossRef] [PubMed]
4. Khan, Z.; Shafait, F.; Mian, A. Converting a Common Low-Cost Document Scanner into a Multispectral Scanner. *Sensors* 2 *19*, 3199. [CrossRef]
5. Huang, Z.; Lin, J.; Yang, H.; Wang, H.; Bai, T.; Liu, Q.; Pang, Y. An Algorithm Based on Text Position Correction and Enco Decoder Network for Text Recognition in the Scene Image of Visual Sensors. *Sensors* **2020**, *20*, 2942. [CrossRef]
6. Nasir, I.M.; Khan, M.A.; Yasmin, M.; Shah, J.H.; Gabryel, M.; Scherer, R.; Damaševičius, R. Pearson Correlation-Based Fea Selection for Document Classification Using Balanced Training. *Sensors* **2020**, *20*, 6793. [CrossRef]
7. Nagaoka, Y.; Miyazaki, T.; Sugaya, Y.; Omachi, S. Text Detection Using Multi-Stage Region Proposal Network Sensitive to Scale. *Sensors* **2021**, *21*, 1232. [CrossRef] [PubMed]
8. Yu, X.; Li, H.; Zhang, Z.; Gan, C. The Optimally Designed Variational Autoencoder Networks for Clustering and Recover Incomplete Multimedia Data. *Sensors* **2019**, *19*, 809. [CrossRef] [PubMed]
9. Araújo, T.; Chagas, P.; Alves, J.; Santos, C.; Sousa Santos, B.; Serique Meiguins, B. A Real-World Approach on the Problem Chart Recognition Using Classification, Detection and Perspective Correction. *Sensors* **2020**, *20*, 4370. [CrossRef] [PubMed]

sensors

MDPI

Article

A Visual Sensing Concept for Robustly Classifying House Types through a Convolutional Neural Network Architecture Involving a Multi-Channel Features Extraction

Vahid Tavakkoli *, Kabeh Mohsenzadegan and Kyandoghere Kyamakya

Institute for Smart Systems Technologies, University Klagenfurt, A9020 Klagenfurt, Austria; kabehmo@edu.aau.at (K.M.); kyandoghere.kyamakya@aau.at (K.K.)

* Correspondence: vtavakko@edu.aau.at; Tel.: +43-463-2700-3540

Received: 14 September 2020; Accepted: 2 October 2020; Published: 5 October 2020

Abstract: The core objective of this paper is to develop and validate a comprehensive visual sensing concept for robustly classifying house types. Previous studies regarding this type of classification show that this type of classification is not simple (i.e., tough) and most classifier models from the related literature have shown a relatively low performance. For finding a suitable model, several similar classification models based on convolutional neural network have been explored. We have found out that adding/involving/extracting better and more complex features result in a significant accuracy related performance improvement. Therefore, a new model taking this finding into consideration has been developed, tested and validated. The model developed is benchmarked with selected state-of-art classification models of relevance for the "house classification" endeavor. The test results obtained in this comprehensive benchmarking clearly demonstrate and validate the effectiveness and the superiority of our here developed deep-learning model. Overall, one notices that our model reaches classification performance figures (accuracy, precision, etc.) which are at least 8% higher (which is extremely significant in the ranges above 90%) than those reached by the previous state-of-the-art methods involved in the conducted comprehensive benchmarking.

Keywords: classification; house architecture type classification; house type classification; convolutional neural networks

1. Introduction

Most visual sensors integrate an image classification related functional bricks. Indeed, image classification is one of the branches of computer vision. Images are classified based on the information abstracted from a series of sequential functional processes, which are preprocessing, segmentation, feature extraction, and finding best matches [1]. Figure 1 roughly illustrates both the input (s) (i.e., an image or some images) and the output of the classifier module. It gets a color image as input and it returns the house-type label, which may be, for example, a bungalow, a villa, a one-family house, etc. Various factors or artefacts in the input images may result in a significant reduction of the classification confidence. Some examples: artifact in image like garden, poor view of image or their neighbor's houses. Worth mentioning is that object classification from visual sensors generated images is a functional brick of high significance in a series of very practical and useful use cases. Some examples of use-cases, just to name a few, are found in real-world robotic applications, such as image/object recognition [2], emotion sensing [3], search and rescue missions, surveillance, remote sensing, and traffic control [4].

Automatically recognizing the architectural type of a building/house from a photo/image of that building has many applications such as an understanding of the historic period, the cultural influence, a market analysis, city planning, and even a support of the price/value estimation of a given building [5–7].

Various candidate known image classification concepts/models can be used for performing this house classification endeavor.

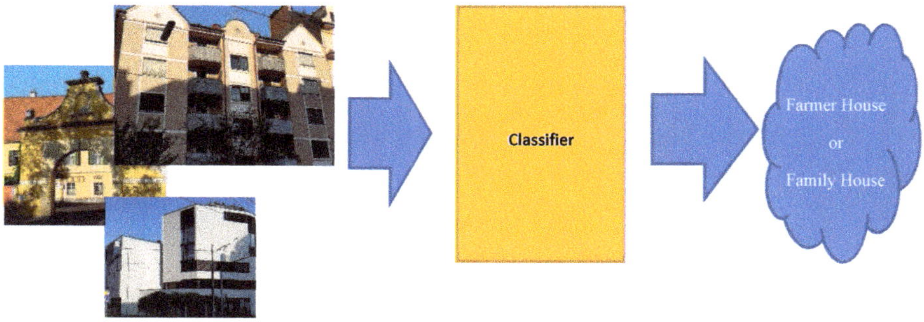

Figure 1. The "House type" classification's overall process pipe (Source: own images).

Thus, as we have a classifier model, the model should be optimized w.r.t. to a related loss function. In this case, we use one of the most famous loss functions, which has been often used for classifications tasks, the so-called categorial cross entropy [8,9]. Equation (1) presents this chosen loss function:

$$L(y, \hat{y}) = -\frac{1}{N}\sum_{j=1}^{M}\sum_{i=1}^{N}\left[y_{i,j}\log\left(\hat{y}_{i,j}\right)\right] \quad (1)$$

where L is the chosen loss function; N is the number of class categories; M is the number of samples; $y_{i,j}$ relates to the different true labels; and $\hat{y}_{i,j}$ relates to the different predicted labels. During the training process, the model will be optimized in a way such that the minimum value of the objective function L in Equation (1) is reached. Subsequently, the model shall be tested and verified.

There are several traditional image classification schemes such as SVM (support vectors machine), just to name one, which can theoretically/potentially be used [10]. However, most of them are not robust enough to capture and learn the relatively very complex patterns involved here in the house classification task although some of them (e.g., SVM) have been proven to be universal approximators. Therefore, one should use/involve truly much high-performing concepts to solve this very difficult/challenging classification task at hand [10]. It is also shown that combining those traditional methods with dynamical neural networks like cellular neural networks can result in a significant performance improvement. For example, Al Machot et al. [11] showed that combining SVM with cellular neural networks considerably improves the SVM performance; this new resulting hybrid model can thus be used as a very fast and robust detector/classifier instead of using the sole SVM model.

In the recent years, the use of convolutional neural networks (CNN) has been increasing at a fast rate for classification and various data processing/mining tasks [12–19]. The input/output data can be represented as arrays or as multi-dimensional data like images. At the heart of a CNN network, we have convolution operators by which the input values in each layer are convoluted with weights matrices [20]. After/before these operations, other operations like sub-sampling (e.g., Max-pooling) or "batch normalization" can be used [17,21]. This process can be repeated and thereby creates several layers of a deep neural network architecture. The last layer is finally connected to a so-called "fully connected" layer. In addition, the network can have some additional channels for different features like putting RGB channels or an edge or blurred image as additional channels [22–26]. The main idea behind this complex structure is based on filtering non-appropriate data. Each filter which is applied will remove some uninteresting/non-appropriate data. Therefore, it results into a smaller network structure and thus the training requires less time as this technique will shrink the searching area.

The Convolutional Neural Network concept was first introduced by Yann LeCun et al. [17] in the 1980s. This model has been created based on both convolutional and sub-sampling layers. Although this model was introduced in the 1980s, it was not yet used popularly in the first years, as computing' processing power and other resources were still very restricted and limited. But nowadays, those restrictions have been removed due to the recent "computing"-related technological advances/progress and one has seen various usages/applications of such neural networks for significantly large problems.

The model developed and used in this paper is based on a CNN architecture, whereby, however, features are extracted through different input channels. In Section 2, we briefly discuss some related works of relevance for house classification. Our novel model is then comprehensively explained in Section 3. Thus, in Section 4, our model is tested and compared with another relevant models while using/involving the very same test data and the results obtained are comprehensively analyzed and discussed. To finish, in Section 5 concluding remarks are summarized.

2. Related Works

Numerous approaches for image classification have been presented over the years. In 1998, LeCun et al. [27] presented a convolutional neural network model to classify handwritten digits. This model (called LeNet-5) comprises three convolutional layers (C1, C3, C5), two average pooling layers (S2, S4), one fully connected layer, and one output layer (see Figure 2). This model involves sigmoid functions to include/consider nonlinearity before a pooling operation. The last layer (see output layers) is using a series of Euclidian Radial Basis Function units (RBF) [28] to classify 10 digits amongst 10 possible classes.

LeNet-5 and LeNet-5-(with distortion) reached after extensive experiments an accuracy of 0.95% and 0.8%, respectively, on the MNIST data set. However, by increasing both the resolution of an image and the number of classes of a classification endeavor, the machine needed for computing consequently requires more powerful processor systems (e.g., GPU units) and a much deeper convolutional neural network model.

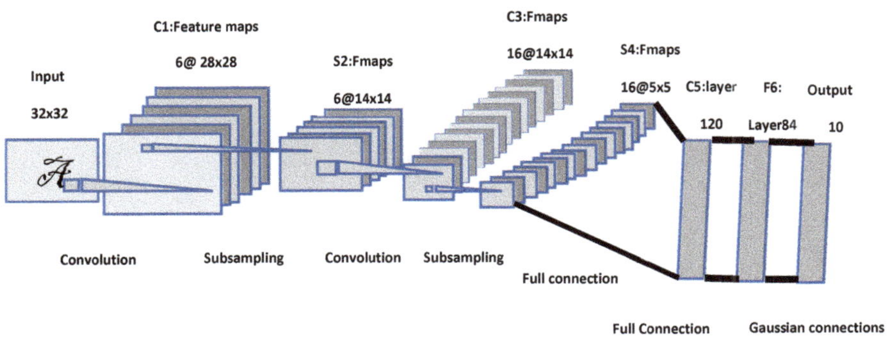

Figure 2. Architecture (our own redrawing) of the LeNet-5 model [27].

In 2006, Geoffery Hinton and Salakhutdinov showed that the neural network with multiple hidden layers can improve the accuracy of classification and prediction by improving different degrees of abstract representation of the original data [29].

In 2012, Krizhevky et al. [30] introduced a large deep CNN (AlexNet). The AlexNet model is much bigger than LeNet-5 with the same acritude (see Figure 3). This model has 5 convolutional layers and 3 fully connected (*FC*) layers. The rectified linear unit (*ReLU*) and the FC layers enables the model to be trained faster than similar networks with *tanh* activation function units. They also added a local response normalization (*LRN*) after the first and the second convolutional layer; that enables the model to normalize information. They further added a max-pooling layer after the fifth convolutional layer and after each *LRN* layer. The stochastic gradient descent (SGD) method has been used for training the AlexNet with a batch size of 128, a weight decay of 0.0005 and a momentum of 0.9. The weight decay works as a regulator to reduce the training error.

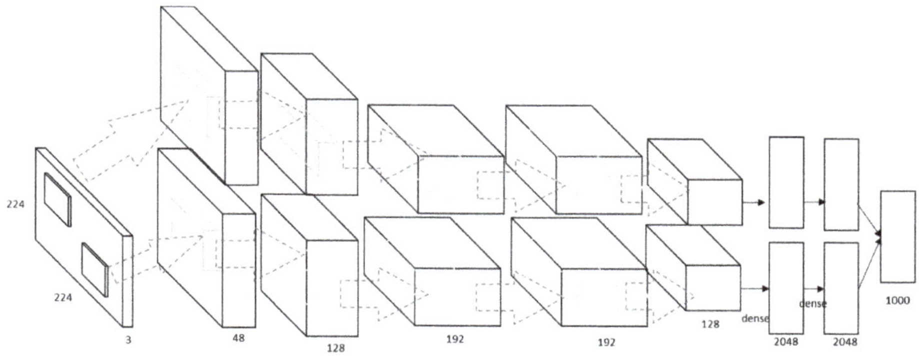

Figure 3. Architecture (our own redrawing) of AlexNet [30].

Also, Jayant et al. [31] presented a model to capture the structural relationships based on statistics of raw-image-patches in different partitions of a document-image. They compared the Relevance Feedback (RF) model to the Support Vector machine (SVM) model and reported that whenever the number of features is large, a combination of SVM and RF is more suitable.

In 2016, He et al. [32] proved that increasing the depth of a CNN processor with more layers increases model complexity on one hand and decrease convergence rate on the other hand. The main problem happens due to introducing new intermediate weights and a consecutive training need to optimize them. For solving this problem, they suggested creating a shallower model with additional layers to perform an identity mapping. Figure 4 shows their core approach.

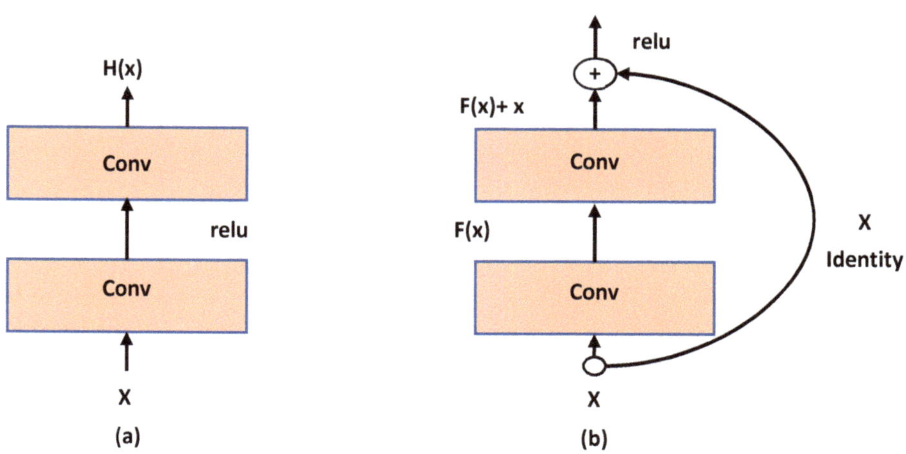

Figure 4. The ResNet model (our own redrawing)—(**a**) Plain layer; (**b**) Residual block [32].

The $H(x)$ block is defined as $H(x) = F(x) + x$. Therefore, $F(x) + x$ will be encapsulated as one block $H(x)$ and the internal complexity of this block shall be hidden. This model is called ResNet and it did show 6% to 9% of accuracy error in classification against the CIFAR-10 test set.

Later, the encapsulation layers concept was extended [33] by introducing a so-called Squeeze-and-Excitation network (SENet). This model reduces the top-5 classification error to 2.25%. The main architecture of this model is shown in Figure 5. Each block is composed of four functions. The first function is a convolution (F_{tr}). The second function is a squeeze function (F_{sq}) which performs an average pooling on each of the channels. The third function is an excitation function (F_{ex}) which is created based on two fully connected neural networks and one activation function (ReLu). The last function is a scale function to generate the final output (F_{scale}). It is known that SENet has shown/demonstrated very good performance results compared to previous models in terms training/testing time and accuracy.

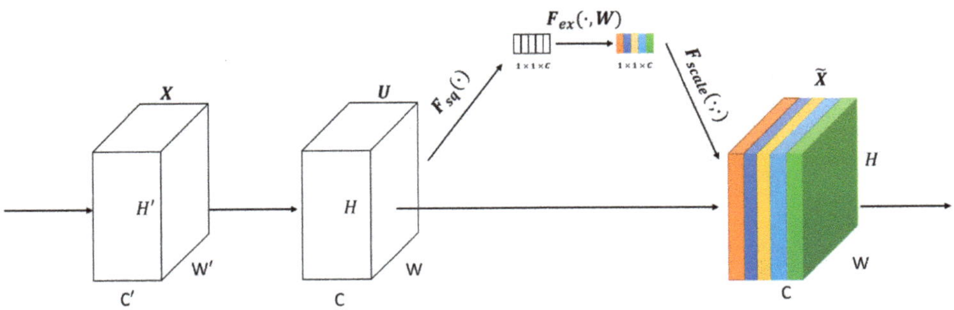

Figure 5. SENet (our own redrawing)—A Squeeze and excitation block [33].

Regarding house classification, a careful study of previous works shows that an automatic detection of architectural styles, and furthers, much harder, even of house types/classes is not yet very well developed/researched [34]. Only few studies on the matter have been published so far. Mathias et al. [35] published a work using SVM to distinguish 4 classes of architectural style, with a specific focus on "inverse procedural modeling"—thereby using imagery to create a generative procedural model for 3D graphics.

Shalunts et al. [36] published a further work to classify the architectural styles of facade windows (see Figure 6). They did thereby use a relatively small dataset (i.e., 400 images) for classifying the architectural styles of buildings through related typical windows in three classes which are: Romanesque, Gothics, and Baroque. Ninety images of the dataset were used for training (i.e., 1/3 of the data of each class).

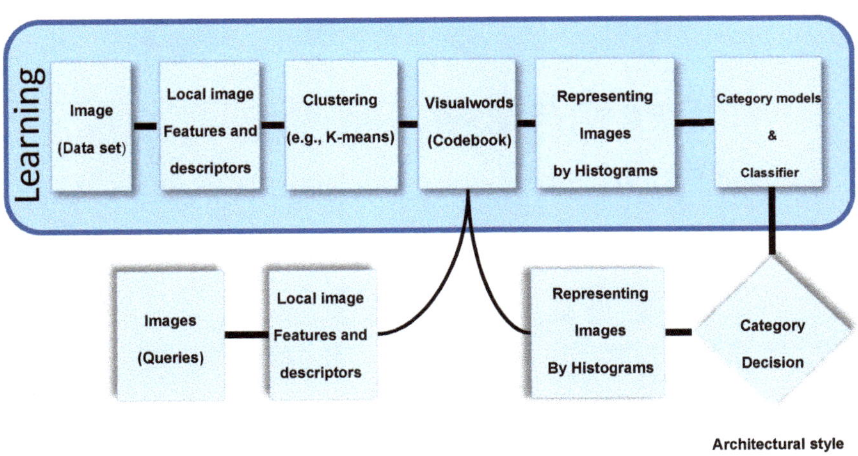

Figure 6. Learning visual words and classification (our own redrawing) scheme [36].

Xu et al. [37] developed a 25-class dataset from Wikimedia and used a model involving HOG that classified through the Multinomial Latent Logistic Regression (see Figure 7). Their model was able to find the presence of multiple styles in the same building through a single image. Notably, they included the "American Craftsman" (one of the house styles used in this work) as a class. Both groups (of authors) lastly mentioned noted the acute absence of a publicly available dataset for architectural style recognition.

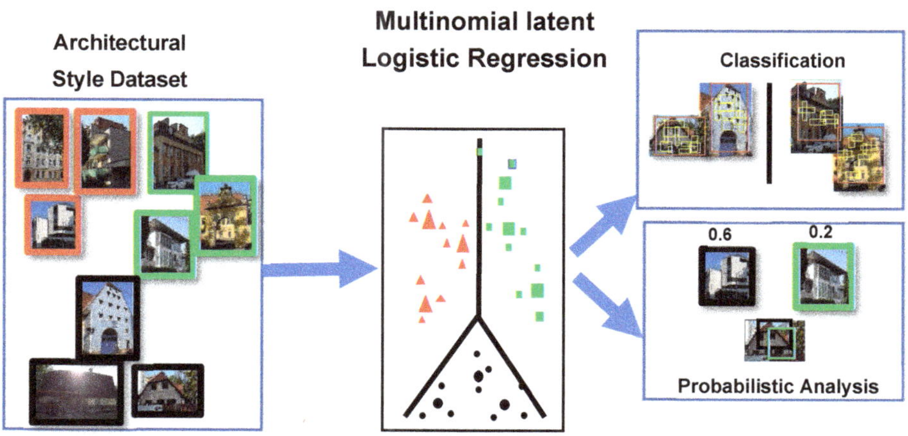

Figure 7. Schematic illustration (our own redrawing) of an architectural style classification using the Multinomial Latent Logistic Regression (MLLR) [37].

In 2015, Lee et al. [38] published a work in which they have used a large dataset of nearly 150 k Google Street View images of Paris, combined with a real estate cadaster map to date building façades and discover the evolution of architectural elements over time (see Figure 8). Their approach used HOG descriptors of image patches to find features correlated with a building's construction time period.

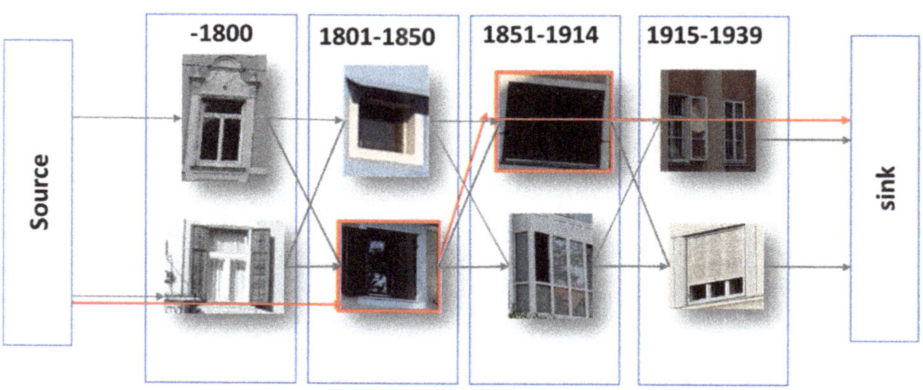

Figure 8. Sample chain graph (our own redrawing). Elements in adjacent periods are fully connected with weights depending on their co-occurrence, while the source and sink connect to every node with weights that penalize the number of skipped periods. Here, the shortest path (in red) skips pre-1800 and 1915–1939 because they lack the long balconies of the other periods. (For clarity, this visualization shows only four periods (instead of ten), and only some source and sink edges [38]).

In 2016, Obeso et al. [39] presented a work based on convolutional neural network (CNN) using sparse features (SF) to classify images of buildings in conjunction with primary color pixel values (see Figure 9). As a result, their mode achieved of 88.01% accuracy.

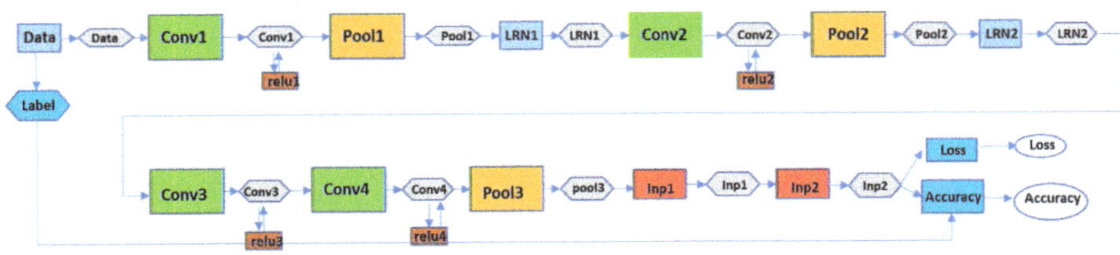

Figure 9. CNN's architecture (our own redrawing), conformed by four convolutional layers, three pooling layers, two normalization layers and two fully-connected layers at the end [39].

We conclude from previous studies that house classification requires very sophisticated classifier models, which shall cover all aspects of the related problem/task and it further becomes evident that CNN is very good candidate for filling this gap (i.e., solving this tough classification task).

3. Our Novel Method/Model

The basic problem formulation has been graphically presented in Figure 10 which essentially underscores the goal of the CNN deep neural model to be developed. However, for reaching the goal with a sufficient accuracy, a series of problems related to the quality of the input "house images" must be solved.

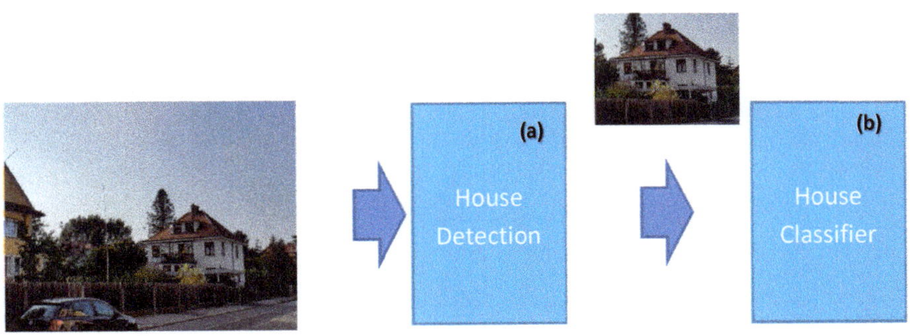

Figure 10. The novel global model is composed of (**a**) house detection and (**b**) classification modules. (Source: our own images).

These problems/issues can be grouped into three different categories (see Figure 11):

- Images, which do not contain a house but only some additional information like garden or trees, make the house classification difficult. Such images are not appropriate for use for a house classification endeavor.
- Some images are (maybe) captured from a very poor angle of the house and thus the house is not well recognizable on them.
- Some house classes have strong similarities with other classes; this is a potential source of misclassification amongst them.

Figure 11. Image problems' illustration: poor view/perspective, more pool garden and/or pool instead of a view of the house, etc. (Source: our own pictures).

For solving the mentioned problems, our overall model (see Figure 11) is designed with two modules: (a) a house detection module, and (b) a house classifier module.

The house detection module is responsible for finding/detecting/localizing a house and its bounding box within the input image. Thus, the result of this module is a bounding box in the input image. It shall also inform us on how much the image has a similarity to a house if at all. This module/layer helps the classifier to perform much better. The second module/layer is for house classification. It may consider all the image or, depending on the outcome of the first module, consider only an image portion within the bounding box identified by the first module/layer. In the lastly mentioned case, the image portion is cropped from the original input image and it becomes the input to be given to the second module for classification.

3.1. House Detection

As explained previously, some images contain either very poor views of the house or/and some additional, for the classifier non-relevant information. Those issues result in decreasing both precision and accuracy of our classifier module. Therefore, this module is responsible for finding the image portion(s) which is/are house views and crop it/them. Figure 11 shows the overall house detection model. The input image is of size 200×200 with three channels. As input images may have different sizes, each original input image must therefore first be rescaled such as to fit either the width or the heights of 200 pixel; the rest of the image may have no values if the image is not a square (i.e., or rectangular form). Therefore, the other parts (with no values) will be black in that case (rectangular form of an original input image). The output of this model (see Figure 12) is one boundary or bounding box. The image portion surrounded by the detected "boundary box" will be cropped out and it will the "input image" for the different classifier models described in Figures 13–15 and the other models involved in the benchmarking process shown in Section 4.

The house detection model contains three main parts: neural layers, feature extraction layers, and a Non-Maximum suppression layer. The feature extraction layers/channels (pre-processors) contain different well-known filters, such as the following ones: Blur filter, Sobel filter, and Gabor filter. These pre-processing filters help/support the model in taking more attention to aspects of the input image which are more important and much relevant. It is the convolutional neural network which is finding the house boundaries. The last part of the CNN architecture is responsible for creating the final boundary boxes by selecting a bounding box with 95% or a higher similarly factor and create the final boundary box based on the Non-Maximum Suppression Algorithm with 0.65 overlap threshold.

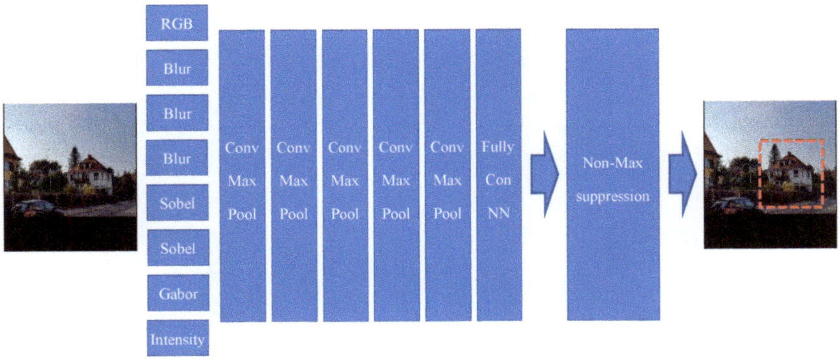

Figure 12. House detection model based on a convolutional neural network. The output of the convolutional neural network will be 4 boundary boxes with four house similarity factors. The boundary boxes with house similarity of 95% will be selected for Non-Maximum suppression with 0.65 overlapping threshold. The house boundary box will be the output of the Non-Maximum suppression module. (source of input image: our own image).

3.2. *House Classification*

The house classification module is designed to classify the input house images into eight different types. Figure 13 shows the overall house classification model. The input image is 200 × 200 with three channels. Cropped images from the previous module are first rescaled to fit either its width or its heights in 200-pixel square, and the rest of the model's input square (of 200 × 200) has no values. Therefore, those rest parts of the input square are black. The output of this model is a class number/label.

On the way to developing the very best model for house classification, we created several models from which to then select the best suitable one for the task at table. These different models are explained in this section.

3.2.1. Model I

Our first classification model is composed of five convolutional layers. The outputs of those convolutional layers go into different max-pool layers. Finally, the output of the last max-pool layer goes into a dense layer, whereby the latest dense layer has eight output neurons, which are representing the eight house classes (see Figure 13).

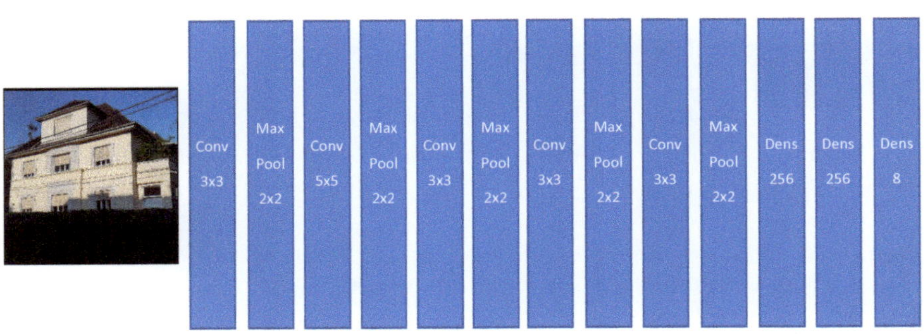

Figure 13. House classification Model I (Source of input image: our own image).

3.2.2. Model II

The second model, like the previous model, has five convolutional layers. The result/output of those convolutional layers will go respectively into max pool layers. Finally, the output of the latest max pool layer will go into the dense layers. The final dense layer has eight output neurons, which represent our eight house classes. The main difference between these two classifier models are the preparation/pre-processing layers of this second model.

These pre-processing layers of this second model provide/generate more details and they are indeed new channels besides the basic the color channels of the input image. These new additional channels are respectively: Blur 3 × 3, Blur 5 × 5, Blur 9 × 9, Sobel Filter X, Sobel Filter Y, and Intensity (see Figure 14).

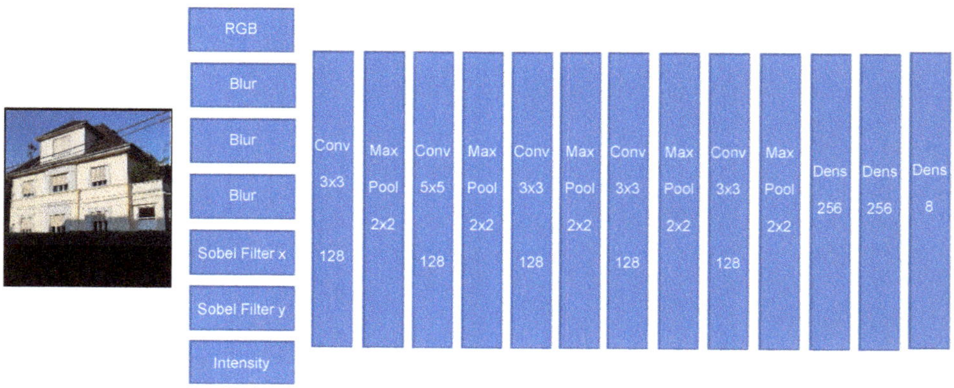

Figure 14. The house classification Model II (Source of input image: our own image).

3.2.3. Model III

This model has also two main parts: a) neural layers, and b) features extraction layers. The features extraction pre-processing layers/channels contain different well-known filters such as the following ones: Blur, Sobel, and Gabor filters (see Figure 15). Here too, these pre-processing filters help/support the model in placing more attention on aspects of the input image, which are more important and relevant for the classification task.

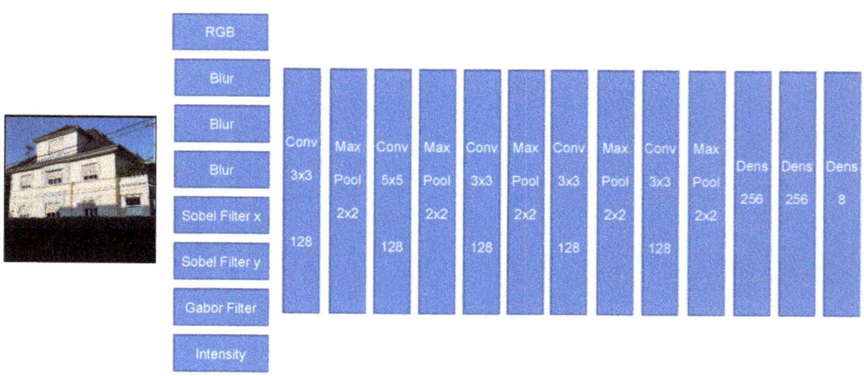

Figure 15. Modell III—Convolutional neural network for house classification. The output of the model consists of 8 house classes (Source of input image: our own image).

Indeed, the pre-processing filters provide more relevant features to the model, and this significantly supports the training process to search and find those features, which are pointing directly to those parts of the input image, which are most relevant. Figure 16 shows, for illustration, the results of the image filtering through one of the pre-processing modules, here the Gabor filters. Each Gabor filtered image is highlighting some interesting features of the image which may help the classifier to better perform the classification task.

Figure 16. Effect of the Gabor filters on an input house image: The top row images are produced by Gabor filters with a kernel size 5, sigma 2, and theta having the following respective values: 0, 45, 90 and 135 degrees (from left to right). The bottom row images are produced by Gabor filters with kernel size 5, sigma 5, and theta having the following respective values: 0, 45, 90 and 135 degrees (from left to right). (Source of input image: own image).

4. Results Obtained and Discussion

As previously explained, several images were gathered from the Internet and used for both training and testing after an appropriate labelling: a total of 1200 images; the number of classes was 8 (see Figure 17 for illustration).

Figure 17. House types which are considered in this work—here some illustrative examples: (**a**) is Farmer house; (**b**) is bungalow; (**c**) is a duplex house; (**d**) is a detached house; (**e**) is an apartment house; (**f**) is a row house; (**g**) is a villa; (**h**) is a country house. (Source of input image: our own images).

The developed deep-learning model (made of two modules: see Figures 12 and 15) was trained with 600 images and verified with 200 images and tested with 400 other images. Figure 14 shows the classification confusion matrix with 200 test images obtained by the best classification model (Figures 12 and 15).

All classifier models have been implemented on a PC with Windows 10 Pro, Intel Core i7 9700K as CPU, double Nvidia GeForce GTX 1080 TI with 8GB RAM as GPU and 64GB RAM. Here, the training takes approximately 8 h.

In order to understand and find an objective justification of why the best model is outperforming the other ones, we conduct a simple feature significance analysis. Hereby, we use the so-called

NMI (normalized mutual information) for the input features. Table 1 shows the Normalized Mutual Information (NMI) scores obtained for the input features. It is clearly shown that by adding more specific features through the multi-channel pre-processing units/modules, the NMI is thereby respectively significantly increased.

Table 1. Normalized Mutual Information (NMI) Scores obtained for the input features for the various deep-learning models used (for the test data sets used in this work).

Model	CNN Model without Multi-Layer Channels (Figure 13)	CNN Model with Multi-Channel Features (Figure 14)	CNN Model with Multi-Channel Features (Figure 15)
NMI	79.5%	84.59%	88.19%

Furthers, Table 2 presents the classification performance scores reached for the three models referred to in Table 1. Here we use the usual multi-dimensional classification performance metrics, namely accuracy, precision, F1-Score, and recall). Most of the classes have an interference/similarity problem with the class "country house"; and it is for this reason often mistaken with other house classes. Therefore, by changing our target function from "Top-1" to "Top-2", our confusion matrix is changed/improved and most of the "similarity" problem is significantly solved/reduced (Figures 18 and 19). Indeed, for practical use cases for which this classification may be relevant (e.g.: assessing the value of a given house for sales or for other purposes), using a "Top-2" classification may be fully sufficient.

Table 2. Comparison of our novel model's classification performance through different traditional metrics.

Model	CNN without Multi-Layer Channels (Figure 13)	CNN with Multi-Channel Features (Figure 15, Top-1)	CNN with Multi-Channel Features (Figure 15, Top-2)
Accuracy	86.5%	94.5%	96.4%
Precision	86.6%	94.2%	96.9%
F1 Score	86.7%	94.0%	96.1%
Recall	87.7%	93.9%	95.9%

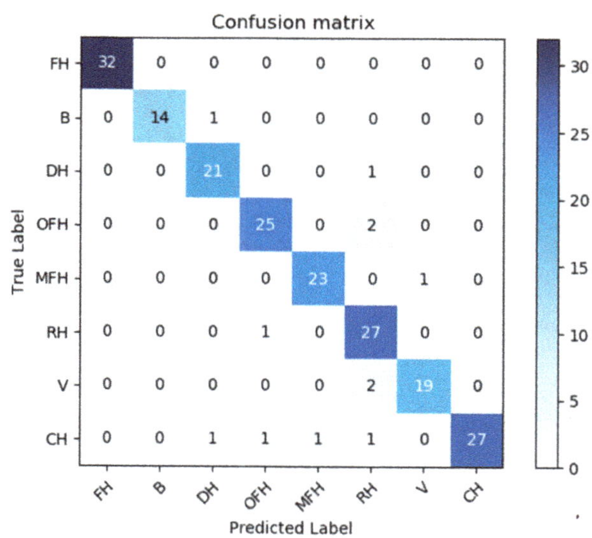

Figure 18. Confusion matrix of the results obtained from Model III while using 200 test images.List of classes: FH is farmer house; B is bungalow house; DH is duplex house; OFH is one family house; MFH is more family house; RH is raw house; V is villa; and CH is country house.

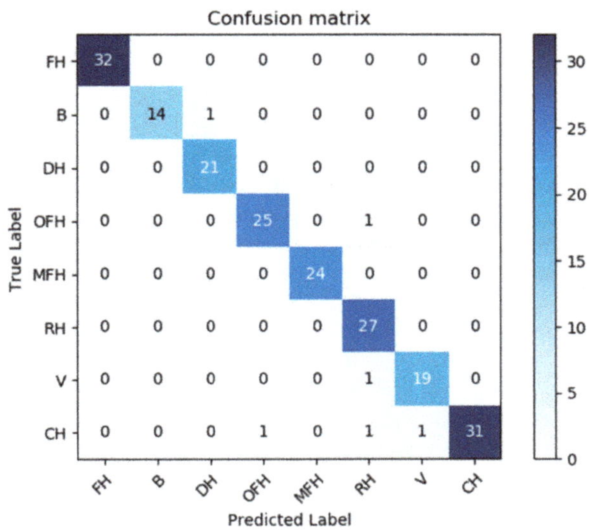

Figure 19. Top-2 Confusion matrix of the results obtained by Model III while using 200 test images. List of classes: FH is farmer house; B is bungalow house; DH is duplex house; OFH is one family house; MFH is more family house; RH is raw house; V is villa; and CH is country house.

In Table 3, the performance of our novel classifier model is compared to that of some very relevant previous/related works. These results clearly show/demonstrate that our novel method (which involves the above discussed multi-channel pre-processing features extraction) has the clearly best performance when compared to the various other models from the relevant recent literature.

Table 3. Comparison of our novel model's performance with that of several other state-of-the-art classifier models published in previous/recent works from the relevant literature.

Model	Mathias (Involving SVM) [35]	Montoya Obesso (Involving CNN) [39]	ResNet-18 [40]	ResNet-34 [40]	CNN without Multi-Layer Channels (Figure 13)	CNN with Multi-Channel Features (Figure 15, Top-1)	CNN with Multi-Channel Features (Figure 15, Top-2)
Accuracy	77.1%	88.1%	78.1%	79.8%	86.5%	94.5%	96.4%
Precision	76.9%	87.7%	78.0%	80.1%	86.6%	94.2%	96.9%
F1-Score	76.5%	87.9%	77.8%	77.8%	86.7%	94.0%	96.1%
Recall	75.3%	88.2%	77.9%	75.8%	87.7%	93.9%	95.9%
Memory Usage	200 MB	100 MB	24 MB	34 MB	20 MB	67 MB	67 MB
Processing Time	100 ms	12 ms	11 ms	12 ms	9 ms	10 ms	10 ms

One can see in Table 3 that our first CNN model without any additional preprocessing is much faster than all other models. However, after adding the pre-processing modules (for additional features) to our first model, the classification performance increases. This can also be seen in Table 1. In addition, both memory and processing time increase after adding the pre-processing layers.

In order to improve the overall classification performance of the housing prediction, the developed model has been divided into two modules: the pre-processors module, and the deep-learning module. The experimental results obtained show that this novel model significantly improves the classification performance. The price is, however, that more memory is consumed (although not very excessive) and the processing time slightly increases.

5. Conclusions

In this paper, a new CNN model for house types classification has been comprehensively developed and successfully validated. Its performance has also been compared to that of some recent very relevant previous works from literature. We can say clearly state that **our novel classification model has a much better performance w.r.t. classification performance** (i.e., accuracy, precision, recall, F1 score), **memory usage, and even, to a large extent, also w.r.t. processing time.**

An objective justification/explanation of the superiority of our novel model presented in Figure 15 is also shown through the fact that adding more features through the different pre-processing units significantly increases the resulting related "NMI scores" metric. Indeed, we thus understand why adding additional features (through Sobel and Gabor filters) has resulted in significantly increasing the model's classification performance (i.e., accuracy, precision, etc.)

Nevertheless, one could observe some misclassifications: a close analysis of the causes of them may inspire future works to reach a much better classification performance. Indeed, the fact of adding several pre-processing features extracting channels in the best-performing version of our novel model has some drawbacks: (a) it uses more memory compared to the (our first) model without those additional pre-processing channels; and (b) the training time is much longer, comparatively.

In addition, a few classification errors have been observed. These misclassifications appear to be caused by the fact that certain house classes/types have a very strong similarity to one another. Examples: class "Villa" and class "Detached house". This requires and inspires some future/further deep investigations and a subsequent better definition of house classes or, as a further option, a merging of some classes, which are visibly too similar to each other. All this does and shall have (in future works) the potential to make the overall resulting classification performance much more accurate and more robust against a series of imperfections of the input house images/photos.

Also, the accuracy of the developed model can be further improved by extending by involving appropriately adapted inspirations involving, amongst others, a series of technical concepts and or paradigms such as the so-called "Adaptive Recognition" [41], "Dynamic Identification" [42], and "Manipulator controls" [43].

Author Contributions: Conceptualization, K.M., V.T. and K.K.; Methodology, K.K.; Software, K.M. and V.T.; Validation, K.M., V.T. and K.K.; Formal Analysis, K.M.; Investigation, K.M. and V.T.; Resources, K.M.; Data Curation, K.M.; Writing-Original Draft Preparation, K.M. and V.T.; Writing-Review & Editing, K.M., V.T. and K.K.; Visualization, K.M. and V.T.; Supervision, K.K.; Project Administration, K.K. All authors have read and agreed to the published version of the manuscript.

Funding: The results of this paper were obtained in the frame of a project funded by UNIQUARE GmbH, Austria (Project Title: Dokumenten-OCR-Analyse und Validierung).

Acknowledgments: We thank the UNIQUARE employees Ralf Pichler, Olaf Bouwmeester und Robert Zupan for their precious contributions and support.

Conflicts of Interest: The authors declare no conflict of interest.

References

1. Luo, F.; Huang, Y.; Tuc, W.; Liu, J. Local manifold sparse model for image classification. *Neurocomputing* **2020**, *382*, 162–173. [CrossRef]
2. Klinger, T.; Rottensteiner, F.; Heipke, C. A Dynamic Bayes Network for visual Pedestrian Tracking. *ISPRS Int. Arch. Photogramm. Remote Sens. Spat. Inf. Sci.* **2014**, *40*, 145–150. [CrossRef]
3. Wang, K.-C. The Feature Extraction Based on Texture Image Information for Emotion Sensing in Speech. *Sensors* **2014**, *14*, 16692–16714. [CrossRef] [PubMed]
4. Yang, M.Y.; Liao, W.; Yang, C.; Cao, Y.; Rosenhahn, B. Security Event Recognition For Visual Surveillance. *ISPRS Ann. Photogramm. Remote Sens. Spat. Inf. Sci.* **2017**, *4*, 19–26.
5. Kang, J.; Körner, M.; Wang, Y.; Taubenböck, H.; Zhu, X.X. Building instance classification using street view images. *ISPRS J. Photogramm. Remote Sens.* **2018**, *145*, 44–59. [CrossRef]

6. Zhang, Y. Optimisation of building detection in satellite images by combining multispectral classification and texture filtering. *ISPRS J. Photogramm. Remote Sens.* **1999**, *54*, 50–60. [CrossRef]
7. Mukhina, K.D.; Visheratin, A.A.; Mbogo, G.; Nasonov, D. Forecasting of the Urban Area State Using Convolutional Neural Networks. In *Fruct Association*; IEEE: Bologna, Italy, 2018.
8. Ho, Y.; Wookey, S. The Real-World-Weight Cross-Entropy Loss Function: Modeling the Costs of Mislabeling. *IEEE Access* **2020**, *8*, 4806–4813. [CrossRef]
9. Cao, J.; Su, Z.; Yu, L.; Chang, D.; Li, X.; Ma, Z. Softmax Cross Entropy Loss with Unbiased Decision Boundary for Image Classification. In *Chinese Automation Congress*; IEEE: Piscataway, NJ, USA, 2018.
10. Peng, Y.; Cai, J.; Wu, T.; Cao, G.; Kwok, N.; Zhou, S.; Peng, Z. A hybrid convolutional neural network for intelligent wear particle classification. *Tribol. Int.* **2019**, *138*, 166–173. [CrossRef]
11. Machot, F.A.; Ali, M.; Mosa, A.H.; Schwarzlmüller, C.; Gutmann, M.; Kyamakya, K. Real-time raindrop detection based on cellular neural networks for ADAS. *Real-Time Raindrop Detect. Based Cell. Neural Netw. ADAS* **2019**, *16*, 931–943. [CrossRef]
12. Gulshan, V.; Peng, L.; Coram, M. Development and validation of a deep learning algorithm for detection of diabetic retinopathy in retinal fundus photographs. *JAMA* **2016**, *316*, 2402–2410. [CrossRef]
13. Esteva, A.; Kuprel, B.; Novoa, R.; Ko, J.; Swetter, S.; Blau, H.; Thrun, S. Dermatologist-level classification of skin cancer with deep neural networks. *Nature* **2017**, *542*, 115–118. [CrossRef] [PubMed]
14. Sharma, N.; Jain, V.; Mishra, A. An Analysis of Convolutional Neural Networks For Image Classification. *Procedia Comput. Sci.* **2018**, *132*, 377–384. [CrossRef]
15. Wang, Q.Z.X. Street view image classification based on convolutional neural network. In *IAEAC*; IEEE: Chongqing, China, 2017.
16. Ahn, J.; Park, J.; Park, D.; Paek, J.; Ko, J. Convolutional neural network-based classification system design with compressed wireless sensor network images. *PLoS ONE* **2018**, *13*, e0196251. [CrossRef] [PubMed]
17. Lee, S.; Chen, T.; Yu, L.; Lai, C. Image Classification Based on the Boost Convolutional Neural Network. *IEEE Access* **2018**, *1*, 1–10. [CrossRef]
18. Xu, X.; Li, W.; Ran, Q.; Du, Q.; Gao, L.; Zhang, B. Multisource Remote Sensing Data Classification Based on Convolutional Neural Network. *IEEE Trans. Geosci. Remote Sens.* **2017**, *99*, 937–949. [CrossRef]
19. Wenhui, Y.; Fan, Y. Lidar Image Classification Based on Convolutional Neural Networks. In Proceedings of the 2017 International Conference on Computer Network, Electronic and Automation (ICCNEA), Xi'an, China, 23–25 September 2017.
20. Shahid, S.; Shahjahan, M. A new approach to image classification by convolutional neural network. In Proceedings of the 2017 3rd International Conference on Electrical Information and Communication Technology (EICT), Khulna, Bangladesh, 7–9 December 2017.
21. Ioffe, S.; Szegedy, C. Batch Normalization: Accelerating Deep Network Training by Reducing Internal Covariate Shift. *arXiv* **2015**, arXiv:1502.03167.
22. Wang, W.; Lu, X.; Song, J.; Chen, C. A two-column convolutional neural network for facial point detection. In *ICPIC*; IEEE: Shanghai, China, 2017; pp. 169–173.
23. Karpathy, A.; Toderici, G.; Shetty, S.; Leung, T.; Sukthankar, R.; Fei-Fei, L. Large-Scale Video Classification with Convolutional Neural Networks. In Proceedings of the IEEE Conference on Computer Vision and Pattern Recognition, Washington, DC, USA, 23–28 June 2014.
24. Hu, Y.; Li, C.; Dan, H.; Yu, W. Gabor Feature Based Convolutional Neural Network for Object Recognition in Natural Scene. In Proceedings of the 2016 3rd International Conference on Information Science and Control Engineering (ICISCE), Beijing, China, 8–10 July 2016.
25. Hosseini, S.; Lee, S.; Kwon, H.; Koo, H.; Cho, N. Age and gender classification using wide convolutional neural network and Gabor filter. In Proceedings of the 2018 International Workshop on Advanced Image Technology (IWAIT), Chiang Mai, Thailand, 7–9 January 2018.
26. Nguyen, V.; Lim, K.; Le, M.; Bui, N. Combination of Gabor Filter and Convolutional Neural Network for Suspicious Mass Classification. In Proceedings of the 2018 22nd International Computer Science and Engineering Conference (ICSEC), Chiang Mai, Thailand, 21–24 November 2018.
27. Lecun, Y.; Bottou, L.; Bengio, Y.; Haffner, P. Gradient-based learning applied to document recognition. *Proc. IEEE* **1998**, *86*, 2278–2324. [CrossRef]
28. Buhmann, M.D. Radial basis functions. *Acta Numer.* **2000**, *9*, 138. [CrossRef]

29. Hinton, G.; Salakhutdinov, R. Reducing the Dimensionality of Data with Neural Networks. *Science* **2006**, *313*, 5786. [CrossRef]
30. Krizhevsky, A.; Sutskever, I.; Hinton, G.E.; Pereira, F.; Burges, C.J.C.; Bottou, L.; Weinberger, K.Q. Imagenet classification with deep convolutional neural networks. *Adv. Neural Inf. Process. Syst.* **2012**, *25*, 1097–1105. [CrossRef]
31. Kumar, J.; Ye, P.; Doermann, D. Learning document structure for retrieval and classification. In Proceedings of the 21st International Conference on Pattern Recognition (ICPR2012), Tsukuba, Japan, 11–15 November 2012.
32. He, K.; Zhang, X.; Ren, J.S.S. Deep residual learning for image recognition. In Proceedings of the IEEE Conference on Computer Vision and Pattern Recognition (CVPR), Las Vegas, NV, USA, 27–30 June 2016.
33. Hu, J.; Shen, L.; Sun, G. Squeeze-and-excitation networks. *arXiv* **2017**, arXiv:1709.01507.
34. Römer, C.; Plümer, L. Identifying Architectural Style in 3D City Models with Support Vector Machines. *Photogramm. Fernerkund. Geoinf.* **2010**, *5*, 371–384. [CrossRef]
35. Mathias, M.; Martinovic, A.; Weissenberg, J.; Haegler, S.; Van Gool, L. Automatic architectural style recognition. *ISPRS-Int. Arch. Photogramm. Remote Sens. Spat. Inf. Sci.* **2011**, *1*, 171–176. [CrossRef]
36. Shalunts, G.; Haxhimusa, Y.; Sablatnig, R. Architectural Style Classification of Building Facade Windows. In *International Symposium on Visual Computing*; Springer: Berlin/Heidelberg, Germany, 2011.
37. Xu, Z.; Zhang, Y.; Tao, D.; Wu, J.; Tsoi, A. Architectural Style Classification Using Multinomial Latent Logistic Regression. In *Computer Vision – ECCV*; Springer: Cham, UK, 2014.
38. Lee, S.; Maisonneuve, N.; Crandal, D.; Efros, A.; Sivic, J. Linking Past to Present: Discovering Style in Two Centuries of Architecture. In Proceedings of the 2015 IEEE International Conference on Computational Photography (ICCP), Houston, TX, USA, 24–26 April 2015. [CrossRef]
39. Obeso, A.M.; Benois-Pineau, J.; Ramirez, A.; Vázquez, M. Architectural style classification of Mexican historical buildings using deep convolutional neural networks and sparse features. *J. Electron. Imaging* **2016**, *26*, 11. [CrossRef]
40. Pesto, C.; Classifying US Houses by Architectural Style Using Convolutional Neural Networks. Stanford University. Available online: http://cs231n.stanford.edu/reports/2017/pdfs/126.pdf (accessed on 10 June 2019).
41. Qi, W.; Su, H.; Aliverti, A. A Smartphone-Based Adaptive Recognition and Real-Time Monitoring System for Human Activities. *IEEE Trans. Hum. Mach. Syst.* **2020**, *50*, 414–423. [CrossRef]
42. Su, H.; Qi, W.; Yang, C.; Sandoval, J.; Ferrigno, G.; de Momi, E. Deep neural network approach in robot tool dynamics identification for bilateral teleoperation. *IEEE Robot. Autom. Lett.* **2020**, *5*, 2943–2949. [CrossRef]
43. Su, H.; Hu, Y.; Karimi, H.R.; Knoll, A.; Ferrigno, G.; de Momi, E. Improved recurrent neural network-based manipulator control with remote center of motion constraints: Experimental results. *Neural Netw.* **2020**, *131*, 291–299. [CrossRef]

sensors

Article

A New Filtering System for Using a Consumer Depth Camera at Close Range

Yuanxing Dai [1,*], Yanming Fu [2], Baichun Li [3], Xuewei Zhang [1], Tianbiao Yu [1] and Wanshan Wang [1]

1 School of Mechanical Engineering and Automation, Northeastern University, Shenyang 110004, China
2 Laboratory Management Center, Shenyang Sport University, Shenyang 110102, China
3 College of Aeronautical Engineering, Civil Aviation University of China, Tianjin 300000, China
* Correspondence: yuanxing_dai@163.com; Tel.: +86-1662-088-7776

Received: 16 June 2019; Accepted: 1 August 2019; Published: 8 August 2019

Abstract: Using consumer depth cameras at close range yields a higher surface resolution of the object, but this makes more serious noises. This form of noise tends to be located at or on the edge of the realistic surface over a large area, which is an obstacle for real-time applications that do not rely on point cloud post-processing. In order to fill this gap, by analyzing the noise region based on position and shape, we proposed a composite filtering system for using consumer depth cameras at close range. The system consists of three main modules that are used to eliminate different types of noise areas. Taking the human hand depth image as an example, the proposed filtering system can eliminate most of the noise areas. All algorithms in the system are not based on window smoothing and are accelerated by the GPU. By using Kinect v2 and SR300, a large number of contrast experiments show that the system can get good results and has extremely high real-time performance, which can be used as a pre-step for real-time human-computer interaction, real-time 3D reconstruction, and further filtering.

Keywords: depth image filtering; point clouds filtering; Kinect v2; depth resolution; close range; hand pose

1. Introduction

The reasons for the success of consumer depth cameras are low price, acceptable accuracy, lower learning costs, extensive applicability, and excellent portability. It has been applied in the fields such as body and facial recognition, 3D motion capture, and has been developing very rapidly.

Most of the depth cameras are based on time-of-flight principle, such as Kinect v2 and SR300. It can collect the laser spots array reflected by surfaces, and works out the time difference between emission and reflection to get the distance array of the scene [1,2]. Generally, the array is expressed as a gray image, and the gray value of the pixel is generated by the depth value of the position according to certain rules.

According to the principle of perspective, within the unit area, the closer the surface is to the camera, the more laser points will be reflected, which means that higher measurement point density will result in higher surface resolution. Although the range could be changed by using Draelos's method [3], according to our observation, when a target surface gets close to the nearest limit, for example, 3D reconstruction of small objects [1], in the point cloud acquired by consumer depth cameras, there will be some irregular shape of noise areas surrounding or on the edge of the realistic surface (a surface consisting of laser points reflected by a real object, it is used to distinguish unrealistic surfaces formed by noise points that should not exist).

In the process of generating point clouds with depth cameras, we found that the closer the distance is, the more significant the phenomenon is. Figure 1 shows this phenomenon by using the depth images of a human hand at different distances. In static applications, these low confidence noise areas could be filtered effectively in a post-processing stage [4]. However, any time consumption is undesirable for real-time interactive applications [5–7].

Using a consumer depth camera at close range is a double-edged sword. Developers often aim to use depth images or video stream for further development, such as reverse engineering [8], human pose recognition [9–12], and 3D scene reconstruction [13]. In order to obtain a pure depth image with high accuracy and low noise, one option is to select expensive, high learning cost, and precise optical equipment (3D time-of-flight (ToF) camera or LIDAR). However, for time and money savings, an easier way is to place the object closer to get a higher resolution on the surface of the object. In this case, how to eliminate the noise deterioration caused by close-range use has become a problem that must be solved.

Traditional methods for reducing or eliminating color image noise are usually based on window smoothing or sharpening, such as median filtering [14], non-local means filter [15], bilateral filtering [14,16], joint bilateral filtering [17], etc. The principle is to make a window for each pixel in the image, and update the center pixel according to the value of other pixels in the window.

Different filters use different window selection methods and updated strategies [18]. However, the unmodified algorithm transplantation is not very suitable for depth image filtering. For edge noise of a depth image, joint bilateral filtering with reliable sources (usually from color images) can perform very well. It can better preserve the edge details of an object [19], but for human hand depth image filtering it also involves the lighting conditions [20–22] and the color difference of the foreground and background [23,24].

To fill this gap, in this article we proposed a composite filtering system for eliminating low confidence noise areas around or on the realistic surface and obtaining relatively pure point clouds of a human hand within close distance. In order to maximize the retention of raw data for further use, the system does not use a smoothing filtering algorithm. All the algorithms in the system are implemented by GPU-assisted parallel computing, thus making the system achieve very high real-time performance. Finally, the experiment results show that the proposed filtering system could eliminate the vast majority of noise areas.

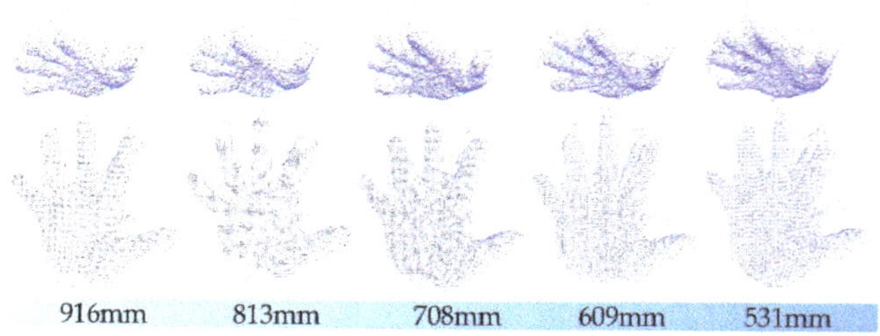

Figure 1. Point clouds of the human hand at different distances. The distance values are obtained by calculating the average of 25 × 25 pixels depth at the center of the hand.

2. Noise Characterization

Severely distorted noise of a close-range collected depth image tends to concentrate on the area of the image where the depth gradient is large. For a highly integrated device, the calibration method for laser scanner cannot be used [25], and as a result, the user cannot adjust the data generation process in most cases (it depends on the camera, SR300 could be allowed to adjust laser intensity and type of built-in filters), but only using the deep data acquired from the device. Hence an in-depth understanding of the noise characteristics within the depth images is the first task to build an effective filtering system.

The noise in the depth image is actually the sum of spatial noise and temporal noise. The former can be construed as an inaccurate depth measurement, which mainly includes the zero depth (ZD) that cannot be measured (like NaN [26]), and the wrong depth (WD) that is far from the actual depth value. The latter refers to the phenomenon that the measured value fluctuates with time when the depth of this point does not change, thus, multi-frames are needed to eliminate the temporal noise [27] that may cause input delay to the interface system. More detailed elaborations are made in [26–29].

For real-time interactive applications, any delay should be avoided, then the best way is to start with spatial noise and eliminate the low confidence WD areas. Therefore, in this section, according to shape and location, the noise areas that seriously affect the correctness of point cloud generation are of the hand surface classified into three types. Two of them are original noise which are shown in Figure 2, and another one is residual noise which will be described in Section 3.3.

- Outlier noise. This is the point with WD that exists away from the realistic surface, and is usually randomly distributed in the depth image spatially and temporally, which usually has impacts on the pass-through filtering [30].
- Edge noise. This kind of noise exists regionally. It can be an unrealistic surface composed of noise points surrounds the realistic surface, or it can be part of the edge of a realistic surface with WD. The closer to the blank area, the greater the depth gradient. It would eventually point to the z_+ or z_- (the positive or negative direction of the depth value).
- Plaque noise. This is a kind of residual noise. The filtering system may miss some plaque areas after filtering the first two kinds of noise. Most of the residual plaque areas are isolated and a few are connected to realistic surfaces.

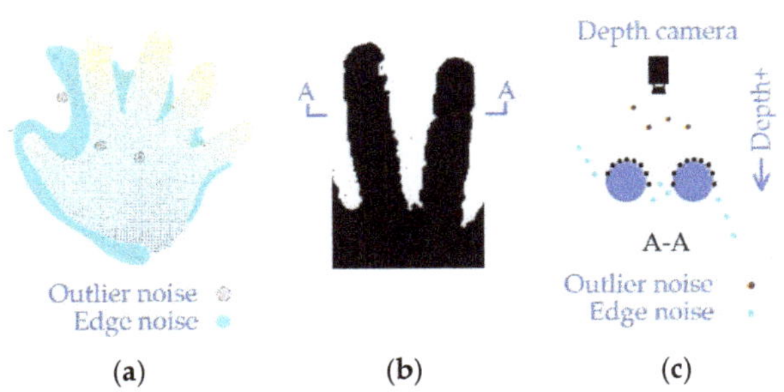

Figure 2. Noise classification. (**a**) A noisy point cloud image of the human hand at a distance of 800–900 mm. In order to show the details of the noise area, position A in (**b**) is cut and the section view is displayed in (**c**).

Together, these three types of noise constitute a low confidence noise area in depth images. It is worth noting that the characteristics of the first two noises are not independent of each other, if the gradient in the window is too large, the extreme points in the same window could be regarded as outliers. Therefore, in the next section, a composite filtering system will be proposed for these types of noise.

3. Proposed Filtering System

Based on the analysis of the noise types in the depth image acquired in close range, different noise characteristics make it difficult for a single filter to perform well. Therefore, a filtering system consisting of multiple detection modules is proposed in Figure 3. The CPU only needs to obtain the depth image from the device, and obtain the filtered point cloud data from the GPU and display them. All filtering algorithms are running in GPU, the calculation part such as standard deviation (SD) calculation, depth to 3D coordinate conversion follows the calculate-when-using principle to reduce the read and write frequency of the graphics card memory. By setting a reasonable number of loops n_{loop}, it could eliminate most of the low confidence areas, and preserve realistic surfaces. Highly parallelized algorithms in the filtering system could save the computing resources of CPU and make the system run in real time.

3.1. Improved Dixon Test

The outliers seriously interfere with the depth value-based hand region truncation, often causing truncation failure. As for the adjacent points belonging to a same realistic surface, their depth values are usually very close. For any non-zero point (NZP) $p \in I_D$, both value range and the SD of $\delta(p)$ could not be too large, where $\delta(p) \subset I_D$ is the neighbor set of p.

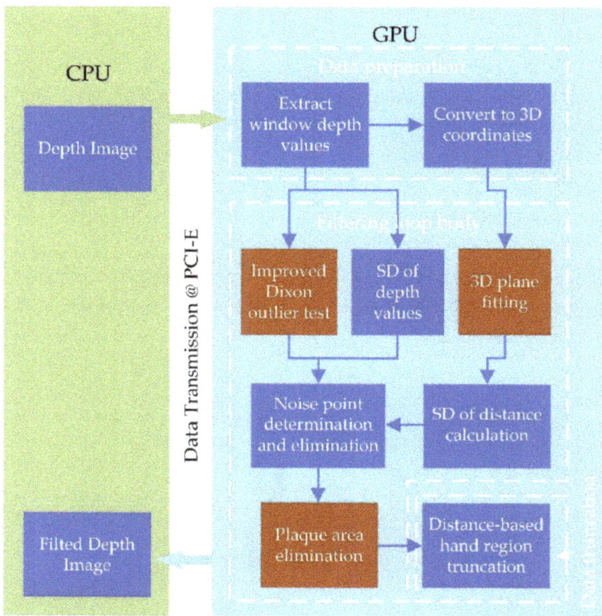

Figure 3. Proposed filtering system for hand depth image. The part with GPU algorithms of the system can be divided into three sub-parts. (1) Data preparation: extract the depth data of all points in the window corresponding to each thread, and work out their 3D coordinates. (2) Filtering loop body: as the core part of the filtering system, its main function is to identify the noise points and filter them out. (3) Data truncation: preserving the foreground and eliminating the depth image of the background. In addition, the three filtering algorithms proposed for different types of noise areas are marked in red boxes.

The central idea of the Dixon test is to determine whether extreme points are outliers by calculating the ratio between extreme point deviation and sample range. Equation (1) shows the Dixon outlier test method for up to 10 samples [31], where the Q_u and Q_l are used to identify the maximum and minimum sample respectively. x_1 and x_n are the extremes of the arranged samples, and the values of Q_u and Q_l could reflect how large the gap is. Different confidence levels (α) correspond to different limits of Q_u and Q_l, it can be obtained by looking up the table [31].

$$\begin{cases} 3 < n < 7 : Q_u = \frac{x_n - x_{n-1}}{x_n - x_1},\ Q_l = \frac{x_2 - x_1}{x_n - x_1} \\ 8 \le n \le 10 : Q_u = \frac{x_n - x_{n-2}}{x_n - x_2},\ Q_l = \frac{x_2 - x_1}{x_{n-1} - x_1} \end{cases},\ x_1, \ldots, x_n \in \delta(p) \tag{1}$$

However, to some extent, the ratio only reflects the deviation between the extreme point and cluster of other points spatially. Therefore, there are natural defects in applying it directly to outlier detection in depth images. The reason is that it cannot reflect the discreteness of depth values of all points in the window macroscopically. However, the SD, which can reflect the dispersion on the value of the samples, cannot microscopically reflect the positional relationship between each point and the cluster. Figure 4 and Algorithm 1 show the improvements of the Dixon test. There are three ways to determine whether p is an outlier.

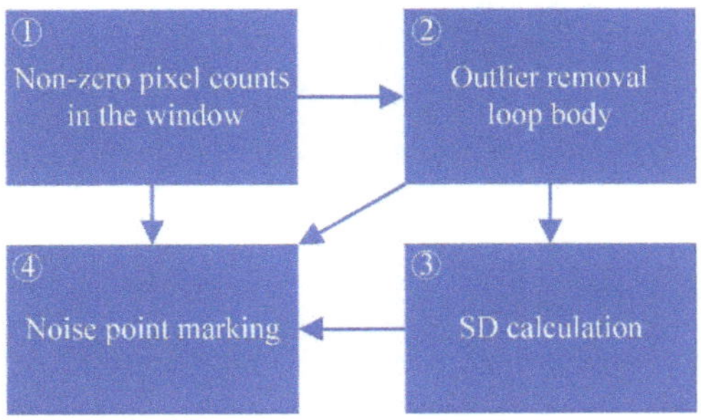

Figure 4. Improved Dixon test for depth image outlier detection. This algorithm combines the continuously draining outliers Dixon test and standard deviation (SD) calculation to perform macroscopic and microscopic identification of the center point.

Algorithm 1 Outliers Detection

Input: $\delta(p)$ **Output:** *Stat(tid)*, true for noise point

count NZPs: $n_{NZP} \leftarrow$ counter($\delta(p)$)
if $n_{NZP} < k$ **then**
    return: *Stat (tid)* $\leftarrow$ *true*
endif
ascending ordering: $H \leftarrow$ sorter($\delta(p)$)
nremain $\leftarrow n_{NZP}$
do
    remove outliers: DixonTest(H, n_{remain})
while $k <$ *nremain* $< n$
if $p \in H$ then
    return: *Stat (tid)* $\leftarrow$ *true*
endif
calculate SD: SD $\leftarrow$ deviation(H)
if $SD > SD_{max}$ **then**
    return: *Stat (tid)* $\leftarrow$ *true*
else
    return: *Stat (tid)* $\leftarrow$ *false*
endif

- ①→④, for any p-centred $k \times k$ size window, when the number of NZP $n_{NZP}(\delta(p)) < k$, that is, there are only at most $k-2$ NZPs except p. As is shown in Figure 5, as an outlier, p locates on the jagged edges or in the blank area, and it can be directly identified as a noise point.
- ① → ② →④, the loop body in ② eliminates the outliers in the sample repeatedly until there are no more points that can be identified as an outlier. If p is eliminated, then it could be defined as an outlier.
- ①→ ② → ③ →④, the remaining NZPs in ② will be sent to an SD calculator to make a macro evaluation of dispersion. p will be marked as an outlier if the SD is too large.

The improved algorithm not only checks whether the extreme point is an outlier, but the SD test is also carried out on the set after eliminating all outliers at the same time, which limits the dispersion of the cluster formed by remaining points. Thus, the macroscopic dispersion and the microscopic position distribution in the window could be simultaneously detected.

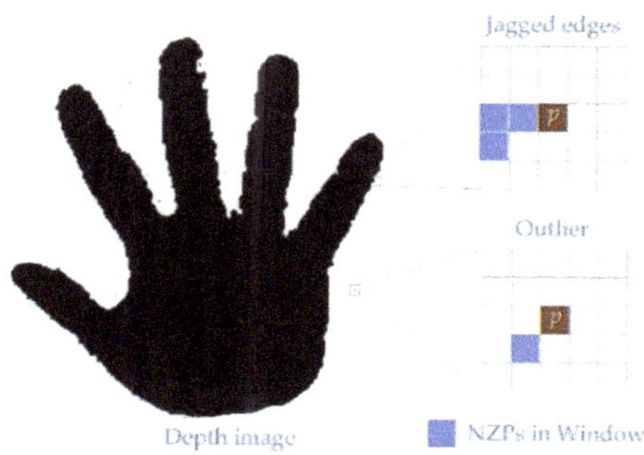

Figure 5. The locations of the points determined to be eliminated. For window areas with a small number of non-zero points, most of them are outliers or located on the jagged edge. This figure shows this case with $k = 5$. Other sizes of windows are similar.

3.2. Edge Noise Filtering Approach

For depth images, the maximum depth gradient within a window should be limited by the line of sight. Thus, if any p is determined as an edge noise point, the maximum gradient in any p-centered window should be larger than a specific value.

In addition, this type of noise area can be eliminated by the approach which is shown in Figure 6. According to $n_{NZP}(\delta(p))$, the determination of edge noise point will be based on the following three cases.

- $n_{NZP}(\delta(p)) < k$, when p locates at the jagged edges, p is defined as a noise point.
- $1.n_{NZP}(\delta(p)) = k$, if NZPs is arranged as a straight-line, p is defined as a noise point.
- $n_{NZP}(\delta(p)) > k$, in this case, fitting a plane π to $G(\delta(p))$ by using the least-square method, where $G(*)$ is the converted 3D global coordinates of $\delta(p)$. Let $S(G(\delta(p)), \pi)$ be the SD of the vertical distances from $G(\delta(p))$ to π, and $\alpha(\pi, \lambda)$ be the angle between π and sight plane λ. If $S(G(\delta(p)), \pi) > S_\pi$ or $\alpha(\pi, \lambda) > \alpha_p$, define p as a noise point.

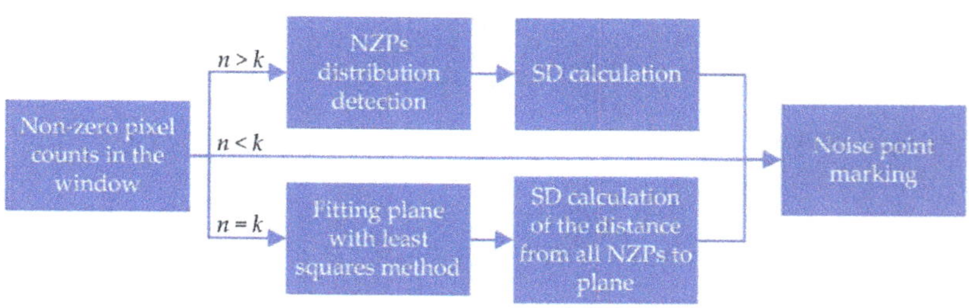

Figure 6. Edge noise filtering. Identification of this kind of noise points is mainly based on the relative position between non-zero points (NZPs).

When $n_{NZP}(\delta(p)) \leq k$, the algorithm makes noise point judgments for the number and distribution of NZPs. In another case, the algorithm fits NZPs into a three-dimensional plane, and the decision is made by the angle of the plane and the dispersion of the distance between the plane and NZPs. s_p and α_p are the given threshold values for judging p's state. Selecting the appropriate threshold will help the system to separate the realistic area from the noise area.

3.3. Plaque Noise Filtering Approach

Part of the plaque noise areas is a kind of residual noise, and they come from the residual part filtered by the above steps. Most of them are isolated and located in the areas that are supposed to be blank as an unrealistic surface. There are also some plaques connected to the realistic surface, which means that the serial algorithm for keeping the hand area by comparing the length of the chain code [32,33] will not always be effective. However, the use of a larger window is likely to cause excessive elimination of the hand area. Therefore, a method for orthogonally detecting the number of consecutive non-zero points is presented in Figure 7 to eliminate the plaque area.

As shown in Figure 7, threads are opened for each $p \in I_D$ to search the adjoining continuous NZPs along rows and columns respectively, and obtain the numbers n_{p_x} and n_{p_y} which stand for how many NZPs are connected to p in the direction of rows and columns (including itself). By putting limits on n_{p_x}, n_{p_y}, and their product m_p, respectively, all plaque areas could be marked. The pseudo-code for one thread is shown in Algorithm 2.

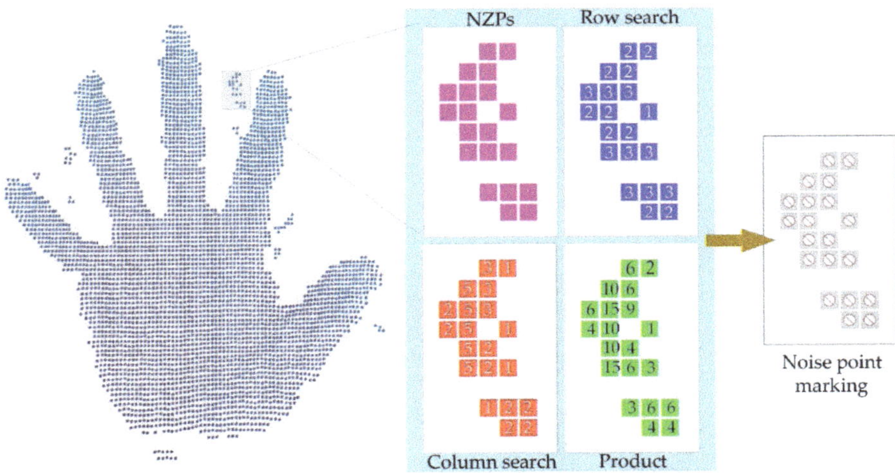

Figure 7. Orthogonally detecting the number of consecutive non-zero points of p.

Algorithm 2 Plaque Noise Area Detection

Input: ID, n_{pymin}, n_{pymin}, m_{pmin}
Output: *Stat(tid)*, true for noise point

row search: Xsearcher (n_{px}, I_D)
col search: Xsearcher (n_{py}, I_D)
$m_p = n_{px} \times n_{py}$
if $n_{px} < n_{pxmin} || n_{py} < n_{pymin} || m_p < m_{pmin}$ **then**
    return: *Stat (tid)* ← true
else
    return: *Stat (tid)* ← false
endif

---

4. Experiments

To the best our knowledge, studies based on a non-smooth filter specially used for filtering the high noise point clouds generated by consuming depth cameras when used at close range have not been reported. Therefore, for comparison, we employed the standard median filter (SMF) and skin color based depth image classification (SCBDIC) (a fused method presented in [34,35]). Its principle is to register the depth and color image, and then remove the unrealistic surface from the point cloud by recognizing the skin color region. However, to obtain universal experimental results, all the experiments were carried out in an indoor fluorescent light environment, and the lighting conditions were not deliberately improved.

All experiments were conducted on a computer with an Intel Core i7 4770 @ 3.6 Ghz CPU and a Nvidia GTX 1060-6 GB graphics card. The depth sequence was captured using a Kinect v2.0 with resolution 512 × 424 at 30 fps and a SR300 with resolution 640 × 480 at 30 fps. The programming environment was Visual Studio 2017 with CUDA 9.2 version.

To prove the validity of our filtering system, we experimented with Kinect v2 and SR300 on hand depth images at different distances. Figure 8 shows the comparison of the proposed filtering system with the other two filters when using Kinect v2. A large amount of edge noise (marked by the blue circle) exists in the original depth image, which constitutes the unrealistic surfaces in the point cloud. However, the part of the color image corresponding to these unrealistic surfaces was not the skin color area, so they could be well removed by SCBDIC. However, the outliers and edge noise located inside the realistic surface (marked by the gray and black circle) could not be filtered out. More seriously, under different light and different angles, the colors had different changes, which can cause many hand areas (marked by the red circle) to be incorrectly recognized, resulting in over-filtering. On the contrary, SMF seems to be ineffective for such large noise areas, and can only filter out some outliers. The change in window size could not provide a better filtering effect, so we present the filtering effect of SMF when $k = 3$ in Figure 8.

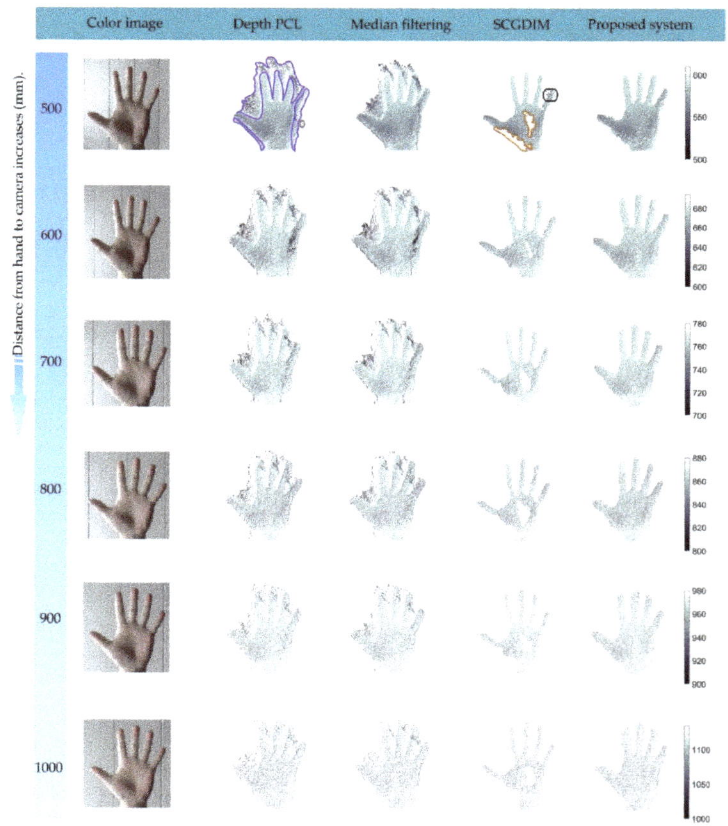

Figure 8. Comparisons at different distances by using Kinect v2.

The proposed filter does not depend on other input sources. As long as the parameters are set reasonably, it can recognize almost all the noise areas and outliers then eliminate them. Even at a close distance of 500 mm, it still produces good results. By querying the judgment status of each sub-filtering algorithm, one type of noise area is not only recognized by its corresponding filtering algorithm, more often, it is recognized by both outlier filtering and edge noise filtering algorithms at the same time. This is because the region recognized as edge noise usually has high SD, which is also one of the characteristics of outlier noise. The reason why p is only recognized as edge noise is that the area where it is located is relatively smooth, but the angle between the fitted plane and the plane of view is too large. In addition, the reason why p is only recognized as an outlier is that in its window, the depth value of p is significantly different from other NZPs, and the values of other NZPs are not much different.

Since point clouds from SR300 hardly produce the realistic surfaces, the SMF that can filter out part of the outliers. Visually, this effect gets better as the filter window increases. As is shown in Figure 9, the edge noise area in orange circle and the outlier in gray circle are smoothed by SMF, and other noise areas in blue circles were also improved to some extent. However, at the same time, the gap between the two fingers (marked in red circle) was filled. At the same time, the depth values of almost all points were changed. This is equivalent to introducing a new error source. The proposed filtering system eliminated almost all edge noise regions without changing the depth value any point and preserved the raw depth data.

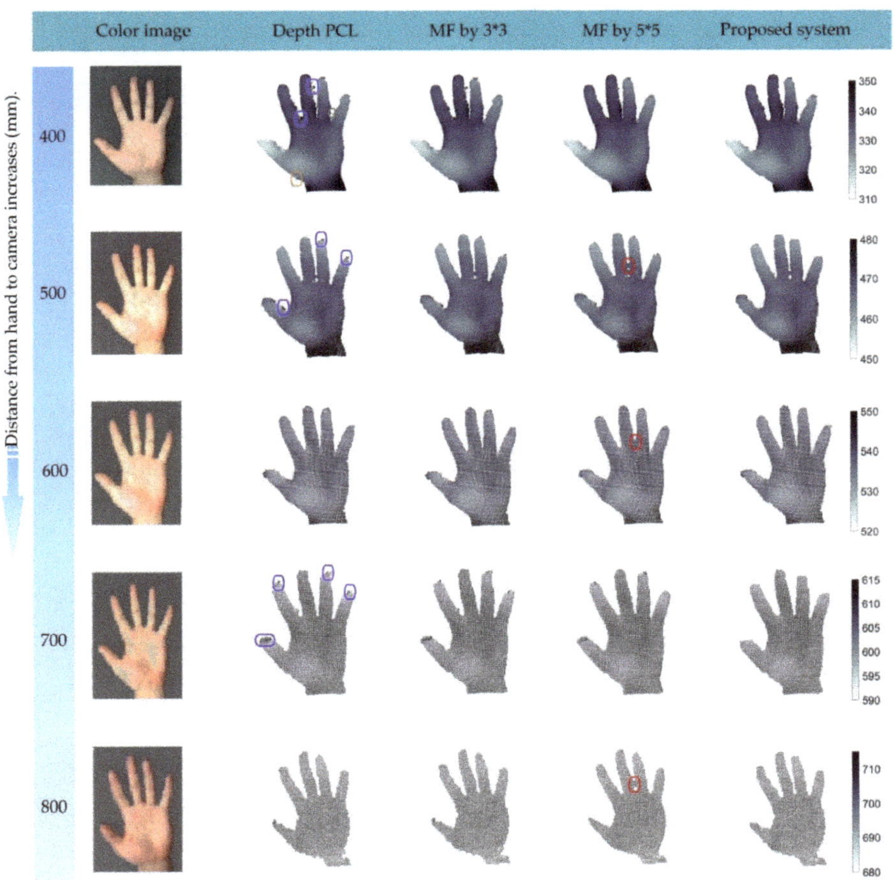

Figure 9. Comparisons at different distances by using SR300.

Figures 10 and 11 show three views of the filtering effect of point clouds by using Kinect v2 and SR300 respectively. For two kinds of equipment with totally different noise characteristics, the proposed system can maintain a good filtering effect. It means that the proposed filtering system has certain universality by setting appropriate parameters. To get better results globally, the determination of the parameters of the filtering system requires a lot of experiments. We present the parameters used in the experiments in this paper which are listed in Table 1.

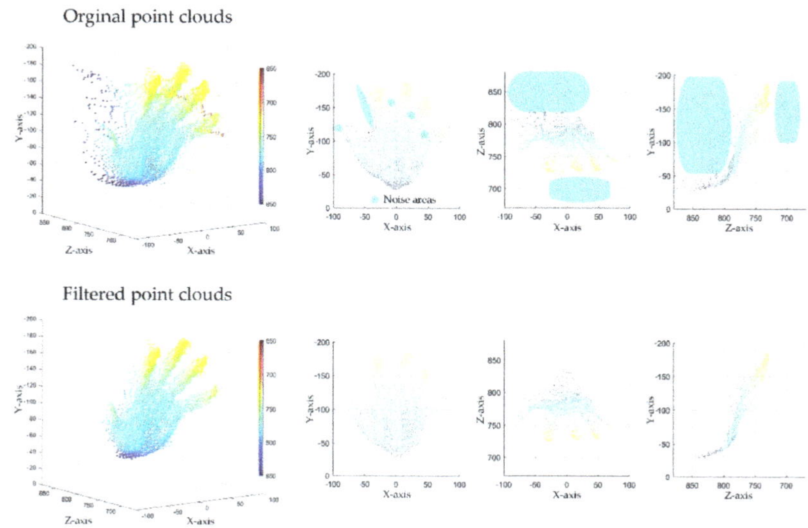

Figure 10. Three views of the filter result (Kinect v2).

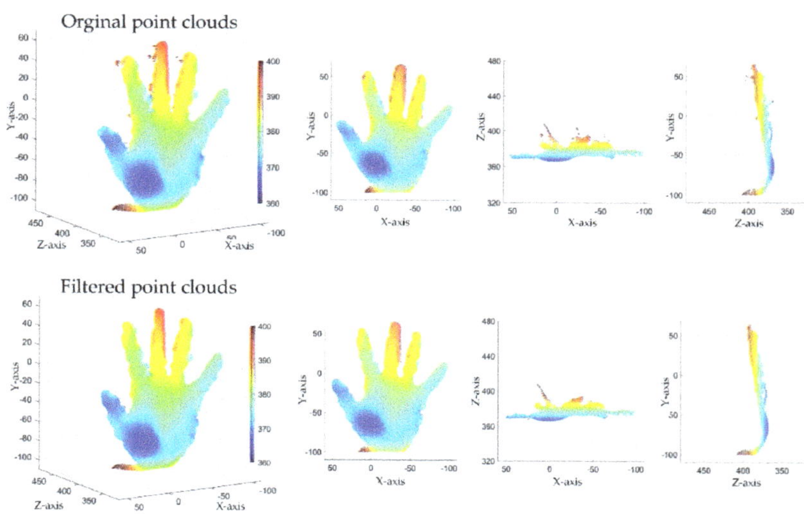

Figure 11. Three views of the filter result (SR300).

To evaluate the stability and real-time performance of the system, 1000 frames of continuous images were recorded as experimental material, and the run time for each frame was stable at 5 milliseconds, and the filtering effect video is updated as the supplementary material. Since all algorithms in this proposed filtering system adopt the parallel structure, the running speed of the system is not very sensitive to the resolution of depth image and more determined by GPU performance.

At the same time, it also has excellent performance in stability. In Figure 12, 18 frames of different hand postures are shown, and most of them have very good filtering effects. However, it is noteworthy that, when a gradient of a part of the realistic surface gets too large, the system may determine it as an edge noise area and eliminate it. The frame in a red box indicates this situation.

Figure 12. Comparisons of different hand gestures.

5. Conclusions

When collecting depth images with a consumer depth camera, the noise interference becomes more serious as the object approaches. In order to eliminate these noise areas and obtain a correct pure raw point cloud with high resolution of the object, we proposed a new filtering system for using consumer depth cameras at close range in this paper.

We classified the noise areas into three types, outlier noise, edge noise, and plaque noise. By analyzing the characteristics of these three noise types, we specially designed a filtering algorithm for each noise type: (1) an improved Dixon test algorithm for filtering outlier noise, (2) a three-dimensional plane fitting method to eliminate edge noise, and (3) an algorithm based on searching for the number of adjacent joints for the plaque noise. All algorithms adopted the parallel structure, which greatly improved the efficiency of the filtering system. The running speed of nearly 200 frames per second can meet the application of most real-time interactive systems. We tested the filtering system using two different depth cameras, and the filtering effects were much better than the other two filters involved in the comparison. This shows that the proposed filtering system has certain universality. At the same time, we also presented the system parameters that can achieve a better global filtering effect with the two cameras. Finally, in order to test the stability of the filtering effect, we used 1000 frame continuous hand depth images as experimental materials. The filtering effects show that the system can effectively eliminate most of the noise areas, and 18 of them were selected to present the filtering effect.

Excellent real-time, good filtering effect, and a certain degree of universality enables the proposed filtering system to be used as a pre-step for real-time human-computer interaction, real-time 3D reconstruction, and further filtering.

Table 1. Experimental parameters.

Dis.\Para.	k	n_{loop}	α	S_p	n_{p_x}	n_{p_y}	m_p	S_π	α_p
				Kinect v2					
500 mm	3	20	0.1000	4.3	7	7	25	3	75
600 mm	3	20	0.1000	4.55	5	5	25	3	75
700 mm	3	15	0.1000	4.8	5	5	25	3	75
800 mm	3	15	0.1000	5.05	4	4	16	3	75
900 mm	3	10	0.1000	5.3	4	4	16	3	75
1000 mm	3	10	0.1000	5.55	4	4	16	3	75
				SR300					
400 mm	3	10	0.1000	2.5	5	5	25	1.5	70
500 mm	3	10	0.1000	2.5	5	5	25	1.5	70
600 mm	3	10	0.1000	2.3	5	5	25	1.2	70
700 mm	3	10	0.1000	2.2	4	4	16	1	70
800 mm	3	10	0.1000	2.2	4	4	16	1	70

Future Works

In the future, on the one hand, we will try to develop a method to evaluate the filtering effect, which can be used to realize the automatic optimization of system parameters, and increase or modify some sub-algorithms by using other kinds of cameras to improve the universality of the filtering system. On the other hand, we will try to develop a new algorithm that replaces the points that are marked as noise instead of simply removing them, making the filtered point cloud image edges smoother.

Supplementary Materials: The following are available online at http://www.mdpi.com/1424-8220/19/16/3460/s1, Video S1: filtering effect at close range by using kinect v2.

Author Contributions: In this study, Y.D. is responsible for literature retrieval, charting, research and design, data collection, data analysis, manuscript writing and other practical work. In the process, we got some suggestions about experimental design from Y.F., adopted some research and design methods from B.L., and got the help of X.Z. in the process of data collection. Finally, after the completion of this paper, it was approved by Professor T.Y. and Professor W.W.

Funding: This research received no external funding.

Conflicts of Interest: The authors declare no conflict of interest.

References

1. Hansard, M.; Horaud, R.; Amat, M.; Evangelidis, G. Automatic detection of calibration grids in time-of-flight images. *Comput. Vision Image Understanding* **2014**, *121*, 108–118. [CrossRef]
2. Grzegorzek, M.; Theobalt, C.; Koch, R.; Kolb, A. Time-of-Flight and Depth Imaging. Sensors, Algorithms, and Applications. *Lect. Notes Comput. Sci.* **2013**, *8200*, 354–360.
3. Draelos, M.; Deshpande, N.; Grant, E. The Kinect up close: Adaptations for short-range imaging. In Proceedings of the Multisensor Fusion & Integration for Intelligent Systems, Hamburg, Germany, 13–15 September 2012.
4. Han, X.-F.; Jin, J.S.; Wang, M.-J.; Jiang, W. Guided 3D point cloud filtering. *Multimedia Tools Appl.* **2018**, *77*, 17397–17411. [CrossRef]
5. Buttazzo, G.; Lipari, G.; Abeni, L.; Caccamo, M. *Soft Real-Time Systems: Predictability vs. Efficiency (Series in Computer Science)*; Plenum Publishing Co.: Pavia, Italy, 2005.
6. Gogouvitis, S.; Konstanteli, K.; Waldschmidt, S.; Kousiouris, G.; Katsaros, G.; Menychtas, A.; Kyriazis, D.; Varvarigou, T. Workflow management for soft real-time interactive applications in virtualized environments. *Future Gener. Comput. Syst.* **2012**, *28*, 193–209. [CrossRef]
7. Ma, Z.; Wu, E. Real-time and robust hand tracking with a single depth camera. *Vis. Comput.* **2014**, *30*, 1133–1144. [CrossRef]
8. Novak-Marcincin, J.; Torok, J. Advanced Methods of Three Dimensional Data Obtaining for Virtual and Augmented Reality. *Adv. Mater. Res.* **2014**, *1025*, 1168–1172. [CrossRef]
9. Nguyen, B.P.; Tay, W.L.; Chui, C.K. Robust Biometric Recognition From Palm Depth Images for Gloved Hands. *IEEE Trans. Hum. -Mach. Syst.* **2015**, *45*, 799–804. [CrossRef]
10. Deng, X.; Yang, S.; Zhang, Y.; Tan, P.; Chang, L.; Wang, H. Hand3D: Hand Pose Estimation using 3D Neural Network. *arXiv* **2017**, arXiv:1704.02224.
11. Sun, X.; Wei, Y.; Liang, S.; Tang, X.; Sun, J. Cascaded hand pose regression. In Proceedings of the IEEE Conference on Computer Vision and Pattern Recognition, Boston, MA, USA, 7–12 June 2015; pp. 824–832.
12. Qian, C.; Sun, X.; Wei, Y.; Tang, X.; Sun, J. Realtime and Robust Hand Tracking from Depth. In Proceedings of the IEEE Conference on Computer Vision and Pattern Recognition, Columbus, OH, USA, 24–27 June 2014; pp. 1106–1113.
13. Morana, M. 3D Scene Reconstruction Using Kinect. In *Advances onto the Internet of Things: How Ontologies Make the Internet of Things Meaningful*; Gaglio, S., Lo Re, G., Eds.; Springer International Publishing: Cham, Switzerland, 2014. [CrossRef]
14. Maimone, A.; Bidwell, J.; Peng, K.; Fuchs, H. Enhanced personal autostereoscopic telepresence system using commodity depth cameras. *Comput. Graph.* **2012**, *36*, 791–807. [CrossRef]
15. Buades, A.; Coll, B.; Morel, J. A non-local algorithm for image denoising. In Proceedings of the 2005 IEEE Computer Society Conference on Computer Vision and Pattern Recognition (CVPR'05), San Diego, CA, USA, 20–25 June 2005; pp. 60–65.
16. Zhang, B.; Allebach, J.P. Adaptive Bilateral Filter for Sharpness Enhancement and Noise Removal. *IEEE Trans. Image Process.* **2008**, *17*, 664–678. [CrossRef]
17. Petschnigg, G.; Szeliski, R.; Agrawala, M.; Cohen, M.; Hoppe, H.; Toyama, K. Digital photography with flash and no-flash image pairs. *ACM Trans. Graph.* **2004**, *23*, 664–672. [CrossRef]
18. Essmaeel, K.; Gallo, L.; Damiani, E.; De Pietro, G.; Dipanda, A. Comparative evaluation of methods for filtering Kinect depth data. *Multimed. Tools Appl.* **2015**, *74*, 7331–7354. [CrossRef]
19. Lo, K.-H.; Wang, Y.-C.F.; Hua, K.-L. Edge-Preserving Depth Map Upsampling by Joint Trilateral Filter. *IEEE Trans. Cybern.* **2018**, *48*, 371–384. [CrossRef] [PubMed]
20. Yuan, L.; Sun, J.; Quan, L.; Shum, H.Y. Image deblurring with blurred/noisy image pairs. *ACM Trans. Graph.* **2007**, *26*, 1. [CrossRef]
21. Le, A.V.; Jung, S.W.; Won, C.S. Directional Joint Bilateral Filter for Depth Images. *Sensors* **2014**, *14*, 11362–11378. [CrossRef]
22. Ran, L.; Li, B.; Huang, Z.; Cao, D.; Tan, Y.; Deng, Z.; Miao, X.; Jia, R.; Tan, W. Hole filling using joint bilateral filtering for moving object segmentation. *J. Electron. Imaging* **2014**, *23*, 063021.
23. Cai, Z.; Han, J.; Liu, L.; Shao, L. RGB-D datasets using microsoft kinect or similar sensors: A survey. *Multimedia Tools Appl.* **2017**, *76*, 4313–4355. [CrossRef]

24. Camplani, M.; Mantecon, T.; Salgado, L. Depth-Color Fusion Strategy for 3-D Scene Modeling With Kinect. *IEEE Trans. Cybern.* **2013**, *43*, 1560–1571. [CrossRef]
25. Reshetyuk, Y. *Terrestrial Laser Scanning: Error Sources, Self-Calibration and Direct Georeferencing*; VDM Verlag: Stockholm, Sweden, 2009.
26. International Organization for Standardization. ISO/IEC/IEEE 60559:2011, Information Technology—Microprocessor Systems—Floating-Point Arithmetic. 2011. Available online: https://www.iso.org/standard/57469.html (accessed on 7 August 2019).
27. Nazir, S.; Rihana, S.; Visvikis, D.; Fayad, H. Technical Note: Kinect V2 surface filtering during gantry motion for radiotherapy applications. *Med. Phys.* **2018**, *45*, 1400–1407. [CrossRef]
28. Yu, Y.; Song, Y.; Zhang, Y.; Wen, S. A Shadow Repair Approach for Kinect Depth Maps. In Proceedings of Computer Vision–ACCV 2012, Proceedings of the 11th Asian Conference on Computer Vision, Daejeon, Korea, 5–9 November 2012.
29. Accuracy (Trueness and Precision) of Measurement Methods and Results—Part 1: General Principles and Definitions. Available online: https://www.iso.org/standard/11833.html (accessed on 31 May 2019).
30. Filtering a PointCloud Using a PassThrough Filter. Available online: http://pointclouds.org/documentation/tutorials/passthrough.php (accessed on 20 May 2019).
31. Böhrer, A. One-sided and Two-sided Critical Values for Dixon's Outlier Test for Sample Sizes up to n = 30. *Econ. Qual. Control* **2008**, *23*, 5–13. [CrossRef]
32. Kim, S.D.; Lee, J.H.; Kim, J.K. A new chain-coding algorithm for binary images using run-length codes. *Comput. Vis. Graph. Image Process.* **1988**, *41*, 114–128. [CrossRef]
33. Structural Analysis and Shape Descriptors. Available online: https://docs.opencv.org/2.4/modules/imgproc/doc/structural_analysis_and_shape_descriptors.html?highlight=findcon#cv2.findContours (accessed on 20 May 2019).
34. Khan, R.; Hanbury, A.; Stöttinger, J.; Bais, A. Color based skin classification. *Pattern Recognit. Lett.* **2012**, *33*, 157–163. [CrossRef]
35. Kang, S.I.; Roh, A.; Hong, H. Using depth and skin color for hand gesture classification. In Proceedings of the IEEE International Conference on Consumer Electronics, Las Vegas, NV, USA, 9–12 January 2011.

sensors
MDPI

Article

Robust Combined Binarization Method of Non-Uniformly Illuminated Document Images for Alphanumerical Character Recognition

Hubert Michalak and Krzysztof Okarma *

Faculty of Electrical Engineering, West Pomeranian University of Technology in Szczecin, 70-313 Szczecin, Poland; michalak.hubert@zut.edu.pl

* Correspondence: okarma@zut.edu.pl

Received: 30 March 2020; Accepted: 19 May 2020; Published: 21 May 2020

Abstract: Image binarization is one of the key operations decreasing the amount of information used in further analysis of image data, significantly influencing the final results. Although in some applications, where well illuminated images may be easily captured, ensuring a high contrast, even a simple global thresholding may be sufficient, there are some more challenging solutions, e.g., based on the analysis of natural images or assuming the presence of some quality degradations, such as in historical document images. Considering the variety of image binarization methods, as well as their different applications and types of images, one cannot expect a single universal thresholding method that would be the best solution for all images. Nevertheless, since one of the most common operations preceded by the binarization is the Optical Character Recognition (OCR), which may also be applied for non-uniformly illuminated images captured by camera sensors mounted in mobile phones, the development of even better binarization methods in view of the maximization of the OCR accuracy is still expected. Therefore, in this paper, the idea of the use of robust combined measures is presented, making it possible to bring together the advantages of various methods, including some recently proposed approaches based on entropy filtering and a multi-layered stack of regions. The experimental results, obtained for a dataset of 176 non-uniformly illuminated document images, referred to as the WEZUT OCR Dataset, confirm the validity and usefulness of the proposed approach, leading to a significant increase of the recognition accuracy.

Keywords: image binarization; optical character recognition; document images; local thresholding; image pre-processing; natural images

1. Introduction

The increasing interest in machine and computer vision methods, recently observed in many areas of industry, is partially caused by the growing availability of relatively inexpensive high quality cameras and the rapid growth of the computational power of affordable devices for everyday use, such as mobile phones, tablets, or notebooks. Their popularity makes it possible to apply some image processing algorithms in many new areas related to automation, robotics, intelligent transportation systems, non-destructive testing and diagnostics, biomedical image analysis, and even agriculture. Some methods, previously applied, e.g., for visual navigation in mobile robotics, may be successfully adopted for new areas, such as automotive solutions, e.g., Advanced Driver-Assistance Systems (ADAS). Nevertheless, such extensions of previously developed methods are not always straightforward, since the analysis of natural images may be much more challenging in comparison to those acquired in fully controlled lighting conditions.

One of the dynamically growing areas of the applications of video technologies based on the use of camera sensors is related to the utilization of Optical Character Recognition (OCR) systems. Some of them include: document image analysis, recognition of the QR codes from natural images [1,2], as well as automatic scanning and digitization of books [3], where additional infrared cameras may also be applied, e.g., supporting the straightening process for the scanned pages. Considering the wide application possibilities of binary image analysis for shape recognition, also in embedded systems with limited computational power and a relatively small amount of memory, a natural direction seems to be their utilization in mobile devices. Since modern smartphones are usually equipped with multi-core processors, some parallel image processing methods may be of great interest as well.

As images acquired by vision sensors in cameras are usually full color photographs, which may be easily converted into grayscale images (if they are not acquired by monochrome sensors directly), the next relevant pre-processing step is their conversion into binary images, significantly decreasing the amount of data used in further shape analysis and character recognition. Nevertheless, for the images captured in uncontrolled lighting conditions, the presence of shadows, local light reflections, illumination gradients, and other background distortions may lead to an irreversible loss of information during the image thresholding, causing many errors in character recognition. Hence, an appropriate binarization of such non-uniformly illuminated images is still a challenging task, similar to degraded historical document images containing many specific distortions.

To face this challenge, many various algorithms have been proposed during recent years, i.e., presented at the Document Image Binarization Competitions (DIBCO) organized during the two most relevant conferences in this field: the International Conference on Document Analysis and Recognition (ICDAR) [4] and the International Conference on Frontiers in Handwriting Recognition (ICFHR) [5]. All competitions have been held with the use of dedicated DIBCO datasets (available at: https://vc.ee.duth.gr/dibco2019/) containing degraded handwritten and machine-printed historical document images together with their binary "ground-truth" (GT) equivalents used for verification of the obtained binarization results.

Since there is no single binarization method that would be perfect for all applications for document images, some initial attempts at the combination of widely known approaches have been made [6], although verified for a relatively small number of test images from earlier DIBCO datasets. Another interesting recent idea is the development of some methods, which should be balanced between the processing time and obtained accuracy, presented during the ICDAR 2019 Time-Quality Document Binarization Competition [7]. Some approaches presented during this competition were also based on the combination of multiple methods, e.g., based on supervised machine learning, including texture features, with the use of the XGBoost classifier and additional morphological post-processing, as well as, e.g., a combination of the Niblack [8] and Wolf [9] methods. Nonetheless, such approaches typically do not focus on document images and OCR applications, considering image binarization as a more general task.

Some attempts at the combination of various methods, also using quite sophisticated approaches, have also been made for the images captured by portable cameras [10–12]. Some of the algorithms have been implemented in PhotoDoc [13], a software toolbox designed to process document images acquired with portable digital cameras integrated with the Tesseract OCR engine. A more comprehensive overview of the analysis methods of text documents acquired by cameras may be found in the survey paper [14].

Nevertheless, in view of potential parallelization of processing, an appropriate combination of some recently proposed binarization methods, also with some previously known algorithms, may lead to relatively fast and accurate results in terms of the OCR accuracy.

Although the most common approaches to the assessment of image binarization are based on the comparison of individual pixels [15,16], it should be noted that not all improperly classified pixels have the same influence on the final recognition results. Obviously, incorrectly classified background pixels located in the neighborhood of characters may be more troublesome than single isolated points in the background. Regardless of the presence of some pixel-based measures, such as, e.g., the pseudo-F-measure or Distance Reciprocal Distortion (DRD) [17], considering the distance of individual pixels from character strokes, their direct application would require not only the presence of the GT images, but also their precise matching with acquired photos. Hence, considering the final results of the character recognition, the assessment of thresholding methods considered in the paper is conducted by the calculation of the number of correctly and incorrectly recognized alphanumerical characters instead of single pixels.

One of the main goals of the conducted experiments is the verification of possible combinations of the recently proposed methods [18–20] with some other algorithms, without a priori training, therefore excluding some recently proposed deep learning approaches due to their memory and hardware requirements. To minimize the direct impact of camera parameters and properties on the characteristics of the obtained image and further processing steps, a Digital Single Lens Reflex (DSLR) camera Nikon N70 is used to acquire the images. The main contributions of the paper are the proposed idea of the combination of some recently proposed image binarization methods, particularly utilizing image entropy filtering and multi-layered stack of regions, based on pixel voting, with additional tuning of some parameters of the selected algorithms, as well as verification for the developed image dataset, containing 176 non-uniformly illuminated document images.

The rest of the paper contains an overview of the most popular image thresholding algorithms, including recently proposed ideas of image pre-processing with entropy filtering [18], background modeling with image resampling [19], and the use of a multi-layered stack of image regions [20], as well as the discussion of the proposed approach, followed by the presentation and analysis of the experimental results and final conclusions.

2. Overview of Image Binarization Algorithms

Image binarization has a relatively long history due to a constant need to decrease the amount of image data, caused earlier by the limitations of displays, the availability of memory, as well as processing speed. The simplest methods of global binarization of grayscale images are based on the choice of a single threshold for all pixels of the image. Instead of the simplest choice of 50% of the dynamic range, the Balanced Histogram Thresholding (BHT) method may be applied [21], where the threshold should be chosen in the lowest part of the histogram's valley. However, this fast and simple method, initially developed for biomedical images, should be applied only for images with bi-modal histograms due to some problems with big tails in the histogram, being useless for unevenly illuminated document images. Kittler and Illingworth proposed an algorithm [22] minimizing the Bayes misclassification error expressed as the solution of the quadratic equation, assuming the normal distribution of the brightness levels for objects and background, further improved by Cho et al. [23] using the model distributions with corrected variance values.

Another global method, regarded as the most popular one for images with bi-model histograms, was proposed by Nobuyuki Otsu [24]. Its idea utilizes the maximization of inter-class variance equivalent to the minimization of the sum of two intra-class variances calculated for two groups of pixels, representing the foreground and background, respectively. A similar approach, although replacing the variance with the histogram's entropy, was proposed by Kapur et al. [25]. Since both methods work properly only for uniformly illuminated images, their modifications utilizing the division of images into regions and combining the obtained local and global thresholds were also considered a few years ago [26].

A more formal analysis of the similarities and differences between some global thresholding methods for bi-modal histogram images, including the iterative selection method proposed by Ridler and Calvard [27], may be found in the paper [28]. Nevertheless, these methods do not perform well for natural images, where the bi-modality of the histogram cannot be ensured. A similar problem may be found applying some other methods developed for binarization of images with unimodal histograms [29,30], which are not typical for document images as well.

An obvious solution of these problems is the use of adaptive binarization methods, where the threshold values are determined locally for each pixel, depending on the local parameters, such as average brightness or local variance. In some cases, semi-adaptive versions of global thresholding may be applied as the region based approaches, where different thresholds may be set for various image fragments. One of exemplary extensions of the classical Otsu's method, referred to as AdOtsu, was proposed by Moghaddam and Cheriet [31], who postulated the use of the additional detection of line heights and stroke widths, as well as the multi-scale background estimation and removal.

The region based thresholding using Otsu's method with Support Vector Machines (SVM) was proposed by Chou et al. [32], whereas another application of SVMs with local features was recently analyzed by Xiong et al. [33]. Some relatively fast region based approaches were proposed recently as well [34,35], leading finally to the idea of the multi-layered stack of regions [20].

Apart from the above-mentioned method proposed by Kapur et al. [25], some entropy based binarization methods may be distinguished as well. Some of them, although less popular than histogram based algorithms, utilize the histogram's entropy [36,37], whereas some other approaches are based on the Tsallis entropy [38] or Shannon entropy with the classification of pixels into text, near-text, and non-text regions [39]. Some earlier algorithms, e.g., developed by Fan et al. [40], were based on the maximization of the 2D temporal entropy or minimization of the two-dimensional entropy [41]. Some more sophisticated ideas employ genetic methods [42] and cross-entropy for color image thresholding, as presented in a recent paper [43]. Another recent idea is the application of image entropy filtering for pre-processing of unevenly illuminated document images [18], which may be applied in conjunction with some other thresholding methods, leading to significant improvement, particularly for some simple methods, such as, e.g., Meanthresh, which is based just on the calculation of the mean intensity of the local neighborhood and setting it as the local threshold value.

Another simple local thresholding method using the midgray value, defined as the average of the minimum and the maximum intensity within the local window, was proposed by Bernsen [44]. Although this method may be considered as relatively old, its modification for blurred and unevenly lit QR codes has been proposed recently [45], based on its combination with the global Otsu's method. A popular adaptive binarization method, available in the MATLAB environment as the adaptthresh function, was proposed by Bradley and Roth [46], who applied the integral image for the calculation of the local mean intensity of the neighborhood, as well as the local median and Gaussian weighted mean in its modified versions. A description of some other applications of integral images for adaptive thresholding may be found in the paper [47].

One of the most widely known extensions of the above simple methods, such as Meanthresh or Bernsen's thresholding, was proposed by Niblack [8], who used the mean local intensity lowered by the local standard deviation multiplied by the constant parameter $k = -0.2$ as the local threshold. The default size of the local sliding window was 3×3 pixels, and therefore, the method was very sensitive to local distortions. A simple, but efficient modification of this algorithm, known as the NICK method, was proposed by Khurshid et al. [48] for brighter images with the additional correction by the average local intensity and the changed parameter $k = -0.1$. One of the most popular extensions of this approach was proposed by Sauvola and Pietikäinen [49], where the additional use of the dynamic range of the standard deviation was applied. The additional modifications of this approach were proposed by Wolf and Jolion [9], who used the normalization of contrast and average intensity, as well as by Feng and Tan [50], using the second larger local window for the computation of the local dynamic range of the standard deviation. The latter approach was relatively slow because of the application of additional median filtration with bilinear interpolation. A multi-scale extension of Sauvola's method was proposed by Lazzara and Géraud [51], whereas the additional pre-processing with the use of the Wiener filter and background estimation was used by Gatos et al. [52], together with noise removal and additional post-processing operations.

Another algorithm, known as the Singh method [53], utilizes integral images for local mean and local mean deviation calculations to increase the speed of computations. One of the most recent methods based on Sauvola's algorithm, referred to as ISauvola, was proposed in the paper [54], where the local image contrast was applied to adjust the method's parameters automatically.

Another modification of Sauvola's method applied to QR codes with an adaptive window size based on lighting conditions was recently presented by He et al. [55], who used an adaptive window size partially inspired by Bernsen's approach. Another recently proposed algorithm, inspired by Sauvola's method, named WANafter the first name of one of its authors [56], focuses on low contrast document images, where the local mean values are replaced by so-called "maximum mean", being in fact the average of the mean and maximum intensity values. Nevertheless, this approach was verified only for the H-DIBCO 2016 dataset, containing 14 handwritten images; hence, it might be less suitable for machine-printed document images and OCR applications.

Some other methods inspired by Niblack's algorithm were also proposed by Kulyukin et al. [57] and by Samorodova and Samorodov [58]. The application of dynamic windows for Niblack's and Sauvola's methods was presented by Bataineh et al. [59], whereas Mysore et al. [60] developed a method useful for binarization of color document images based on the multi-scale mean-shift algorithm. A more detailed overview of adaptive binarization methods based on Niblack's approach, as well as some others, may be found in some recent survey papers [61–66].

Some researchers developed many less popular binarization methods, which were usually relatively slow, and their universality was limited due to some assumptions related to necessary additional operations. For example, an algorithm described by Su et al. [67] utilized a combination of Canny edge filtering and an adaptive image contrast map, whereas Bag and Bhowmick [68] presented a multi-scale adaptive–interpolative method, dedicated for documents with faint characters. Another method based on Canny edge detection was presented by Howe [69], who combined it with the Laplacian operator and graph cut method, leading to an energy minimization approach. An interesting method based on background suppression, although appropriate mainly for uniformly lit document images, was developed by Lu et al. [70], whereas Erol et al. [71] used a generalized approach to background estimation and text localization based on morphological operations for documents acquired by camera sensors from mobile phones. The mathematical morphology was also used in the method presented by Okamoto et al. [72].

An algorithm utilizing median filtering for background estimation was recently proposed by Khitas et al. [73], whereas Otsu's thresholding preceded by the use of curvelet transform was described by Wen et al. [74]. Alternatively, Mitianoudis and Papamarkos [75] presented the idea of using local features with Gaussian mixtures. The use of the non-local means method before the adaptive thresholding was examined by Chen and Wang [76], and the method known as Fast Algorithm for document Image Restoration (FAIR) utilizing rough text localization and likelihood estimation was presented by Lelore and Bouchara [77], who used the obtained super-resolution likelihood image as the input for a simple thresholding. The gradient based method for binarization of medical and document images proposed by Yazid and Arof [78] utilized edge detection with the Prewitt filter for the separation of weak and strong boundary points. However, the presented results were obtained using only the document images from the H-DIBCO 2012 dataset.

Some other recent ideas are the use of variational models [79], fast background estimation based on image resampling [19], as well as the application of independent thresholding of the RGB channels of historical document images [80] with the use of Otsu's method. Nevertheless, the latter method requires the additional training of the decision making block with the use of synthetic images. Due to recent advances of deep learning, some attempts were also made [81,82]; although, such approaches needed relatively large training image datasets, and therefore, their application may be troublesome, especially for mobile devices working in uncontrolled lighting conditions. Another issue is related to their high memory requirements, as well as the necessity of using some modern GPUs, which may be troublesome, e.g., in embedded systems, as well as in some industrial applications.

Recently, some applications of the fuzzy approach to image thresholding were also investigated by Bogatzis and Papadopoulos [83,84], as well as the use of Structural Symmetric Pixels (SSP) proposed by Jia et al. [85,86] (the original implementation of the method available at: https://github.com/FuxiJia/DocumentBinarizationSSP). The idea of this method is based on the assumption that the local threshold should be estimated using only the pixels around strokes whose gradient magnitudes are relatively big and directions are opposite, instead of the whole region.

3. Proposed Method

Apart from the approaches presented during the recent ICDAR [87], some initial attempts at the use of multiple binarization methods were made by Chaki et al. [6], as well as Yoon et al. [88], although the presented results were obtained for a limited number of test images taken from earlier DIBCO datasets or captured images of vehicles' license plates. The idea of the combination of various image binarization based on pixel voting presented in this paper was verified using the 176 non-uniformly illuminated document images containing various kinds of illumination gradients, as well as five common font families, also with additional style modifications (bold, italics, and both of them) and utilized the combination of recently proposed methods with some adaptive binarization algorithms proposed earlier, based on different assumptions. The verification of the obtained results was done with the use of three various OCR engines, calculating the F-measure and OCR accuracy for characters, as well as the Levenshtein distance between two strings, which was defined as the number of character operations needed to convert one string into another. All the images were the photographs of the printed documents containing the well-known Lorem ipsum text acquired in various lighting conditions.

Assuming the parallel execution of three, five, or seven various image binarization algorithms, some differences in the resulting images may be observed, particularly in background areas. Nevertheless, the most significant fragments of document images were located near the characters subjected to further text recognition. The main idea of the proposed method of the voting of pixels being the result of the applications of individual algorithms for the same image was in fact equivalent to the choice of the median value of the obtained binary results (ones and zeros) for the same pixel using three, five, or seven applied methods. Obviously, one might not expect satisfactory results for the use of three similar methods, such as, e.g., Niblack's, Sauvola's, and Wolf's algorithms, but for the approaches based on various assumptions, some of the results may differ significantly, being complementary to each other.

The preliminary choice of binarization methods for combination was made analyzing the performance of individual measures for Bickley Diary, Nabuco (dataset available at: https://dib.cin.ufpe.br/), and individual DIBCO datasets, using the typically used measures based on the comparison of pixels (accuracy, F-measure, DRD, MPM, etc.) reported in some earlier papers. Since these datasets, typically used for general-purpose document image binarization evaluation, do not contain ground-truth text data, the OCR accuracy results calculated for our dataset were additionally used for this purpose. Having found the most appropriate combination of three methods, the two additional methods were added in the second stage only to the best combinations of three methods, and finally, the next two methods were added only to the best such obtained combinations of five methods. The choice of the most appropriate candidate algorithms for the combination was made essentially among the algorithms, which individually led to relatively high OCR accuracy.

Considering this, as well as the complexity of many candidate methods, the combination of two recently proposed algorithms, namely image entropy filtering followed by Otsu's global thresholding described in the paper [18] and the multi-layered stack of regions using 16 layers [20], with NICK adaptive thresholding [48], was proposed. Each of these methods may be considered as relatively fast, in particular assuming potential parallel processing, and based on different operations, as shown in earlier papers.

The application of the stack of regions [20] was based on the calculation of the thresholds for image fragments, where the image was divided into blocks partially overlapping each other; hence, each pixel belonged to different regions shifted from each other according to the specified layer, and the final threshold was selected as the average of the threshold values obtained for all regions to which the pixel belonged for different layers. The local thresholds for each region were calculated in a simplified form as $T = a \cdot mean(X) - b$, where $mean(X)$ is the local average, and the values of the optimized parameters were $a = 0.95$ and $b = -7$, as presented in the paper [20].

The application of the image entropy filtering based method [18] was conducted in a few main steps. The initial operation was the calculation of the local entropy, which could be made using MATLAB's entropyfilt function, assuming a 17×17 pixel neighborhood (obtained after the optimization experiments), followed by its negation for better readability. The obtained entropy map was normalized and initially thresholded using Otsu's method to remove the background information partially. Such an obtained image with segmented text regions was considered as the mask for the background subjected to morphological dilation used to fill the gaps containing the individual characters. The minimum appropriate size of the structuring element was dependent on the font size, and for the images in the test dataset, a 20×20 pixel size was sufficient. Such achieved background estimation was subtracted from the original image, and the negative of the result was subjected to contrast increase and final binarization. Since the above steps caused the equalization of image illumination and the increase of its contrast, various thresholding algorithms may be applied in the last step. Nevertheless, the best results of the further OCR in combination with the other methods were obtained for Otsu's global thresholding applied as the last step of this algorithm.

The algorithm described in the paper [19], used in some of the tested variants, was based on the assumption that a significant decrease of the image size, e.g., using MATLAB's imresize function, caused the loss of text information, preventing mainly the background information, similar to (usually much slower) low-pass filtering. Hence, the combination of downsampling and upsampling using the same kernel may be applied for a fast background estimation. In this paper, the best results were obtained using the scale factor equal to 8 and bilinear interpolation. Such an obtained image was subtracted from the original, and further steps were similar to those used in the previous method: increase of contrast (using the coefficient 0.4), negation, and the final global thresholding using Otsu's method as well. Although both methods were based on similar fundamentals, the results of background estimation using the entropy filtering and image resampling differed significantly; hence, both methods could be considered as complementary to each other.

The last of the methods applied in the proposed approach, known as NICK [48], named after the first letter of its authors' names, was one of the modifications of Niblack's thresholding, where the local threshold is determined as:

$$T = m + k \cdot s = m + k \cdot \sqrt{B}\,, \tag{1}$$

where m is the local average value, $k = -0.2$ is a fixed parameter, s stands for the local standard deviation, and hence, B is the local variance.

The modifications behind the NICK method lead to the formula:

$$T = m + k \cdot \sqrt{B + m^2}\,, \tag{2}$$

with the postulated values of the parameter $k = -0.1$ for the OCR applications. As stated in the paper [48], the application of this value of k left the characters "crispy and unbroken" for the price of the presence of some noisy pixels. The window size originally proposed in the paper [48] was 19×19 pixels; however, the suitable parameters depended on the image size, as well as the font size and may be adjusted for specific documents. Nevertheless, after experimental verification, the optimal choice for the testing dataset used in this paper was a 15×15 pixel window with the "original" Niblack's parameter $k = 0.2$.

Since most of the OCR engines utilized their predefined thresholding methods, which were integrated into the pre-processing procedures, the input images should be binarized prior the use of the OCR software to prevent the impact of their "built-in" thresholding. The well-known commercial ABBYY FineReader uses the adaptive Bradley's method, whereas the freeware Tesseract engine developed by Google after releasing its source code by HP company [89] employs the global Otsu binarization. In this case, forced prior thresholding replaces the internal default methods of the OCR software.

4. Discussion of the Results

The experimental verification of the proposed combined image binarization method for the OCR purposes should be conducted using a database of unevenly illuminated document images, for which the ground truth text data are known. Unfortunately, currently available image databases, such as the DIBCO [4], Bickley Diary [90], or Nabuco datasets [87], used for the performance analysis of image binarization methods contain usually a handwritten text (in some cases, also machine-printed) subjected to some distortions such as ink fading, the presence of some stains, or some other local distortions.

Hence, a dedicated dataset containing 176 document images photographed by a Nikon N70 DSLR camera with a 70 mm focal length with the well-known Lorem ipsum text consisting of 563 words was developed with five font shapes, also with style modifications, and various types of non-uniform illuminations. Since the most popular font shapes were used, namely Arial, Times New Roman, Calibri, Verdana, and Courier, the obtained document images may be considered as representative for typical OCR applications. Three sample images from the dataset are shown in Figure 1. The whole dataset, referred to as the WEZUT OCR Dataset, has been made publicly available and may be accessed free of charge at http://okarma.zut.edu.pl/index.php?id=dataset&L=1.

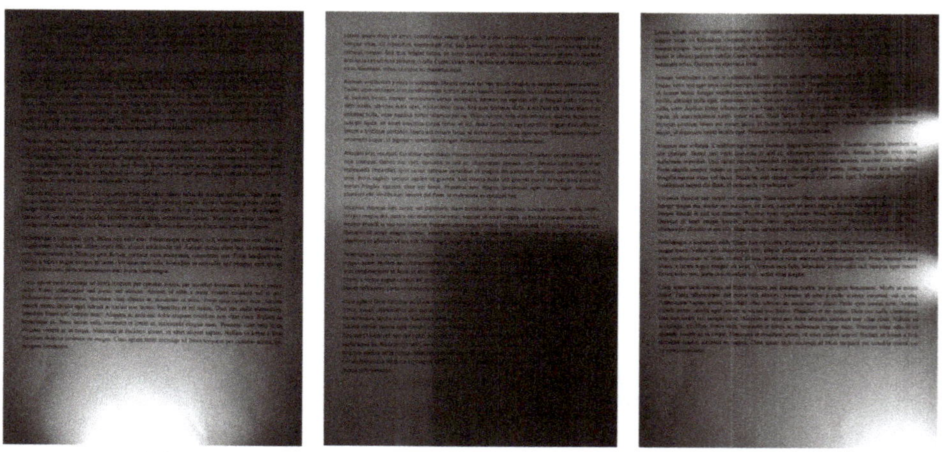

(a) (b) (c)

Figure 1. Three sample unevenly illuminated images from the dataset used in experiments. (**a**) with strongly illuminated bottom part; (**b**) with regular shadows; (**c**) with strongly illuminated right side.

For all images, several image binarization methods were applied, as well as their combinations based on the proposed pixel voting for 3, 5, and 7 methods. Such obtained images were treated as input data for three OCR engines: Tesseract (Version 4 with leptonica-1.76.0), MATLAB's R2018a built-in OCR procedure (also originating from Tesseract), and GNU Ocrad (Version 0.27) based on a feature extraction method (software release available at: https://www.gnu.org/software/ocrad/). Since the availability of some other cloud solutions, usually paid, e.g., provided by Google or Amazon, may be limited in practical applications, we focused on two representative freeware OCR engines and MATLAB's ocr function, which do not utilize any additional text operations related, e.g., to dictionary or semantic analysis.

Each result of the final text recognition was compared with ground truth data (the original Lorem ipsum text) using three measures: Levenshtein distance, interpreted as the minimum number of text changes (insertions, deletions, or substitutions of individual characters) needed to change a text string into another, as well as the F-measure and accuracy, typically used in classification tasks. The F-measure is defined as the harmonic mean of precision (true positives to all/true and false/positives ratio) and recall (ratio of true positives to the sum of true positives and false negatives), whereas accuracy may be calculated as the ratio of the sum of true positives and true negatives to all samples.

To verify the possibilities of the application of various combinations of different methods, the results of the proposed pixel voting approach were obtained using various methods. Nevertheless, only the best results are presented in the paper and compared with the use of individual thresholding methods. Most of the individual methods were implemented in MATLAB, although some of them partially utilized available codes provided in MATLAB Central File Exchange (Jan Motl) and GitHub (Doxa project by Brandon M. Petty). It is worth noting that the initial idea was the combination of three recently proposed approaches described in the papers [18–20]; hence, the first voting (Method No. #37 in Table 1 was used for these three algorithms (similar to the OR and AND operations shown as Methods #35 and #36 in Table 1). Nevertheless, during further experiments, better results were obtained replacing the resampling based method [19] with the NICK algorithm [48]. To illustrate the importance of an appropriate choice of individual methods for the voting procedure, some of the worse results (Methods #39–#41) are presented in Tables 1–3 as well. Further experiments with additional application of some other recent methods led to even better results.

A comparison of the results obtained for the whole dataset using Tesseract OCR is presented in Table 1, together with the rank positions for each of the methods. The overall rank was calculated using the rank positions achieved by each method according to three measures. Method #21 was the modification of Method #20 [18] with the use of the Monte Carlo method to speed up the calculations due to the decrease in the number of analyzed pixels. Nevertheless, applying the integral images in the methods referred to as #14–#20, it was possible to achieve even faster calculations. The results obtained for MATLAB's built-in OCR and GNU Ocrad are presented in Tables 2 and 3, respectively. A comparison of the processing time, relative to Otsu's method, is shown in Table 4. The reference time obtained for Otsu's method using a computer with Core i7-4810MQ processor (four cores/eight threads), 16GB of RAM, and an SSD disk was 1.77 ms.

Analyzing the results provided in Tables 1–3, it may be clearly observed that the best results were achieved using the Tesseract OCR, and the results obtained for the two remaining OCR programs should be considered as supplementary. Particularly poor results could be observed for the GNU Ocrad software. Among the various combinations based on voting, most of them achieved much better results than individual binarization methods regardless of the applied OCR engine, proving the advantages of the proposed approach. Nevertheless, considering the best results, it is worth noting that the use of only three methods (referred to as #58 in Table 1) provided the best F-measure and accuracy and the second results in terms of Levenshtein distance being better even in comparison with the voting approach with the use of five or seven individual algorithms. The Levenshtein distance achieved by this proposed method was only slightly worse than the result of pixel voting using seven algorithms (referred to as #61). Considering the worse OCR engines, some other combinations led to better results, especially for GNU Ocrad, where the application of seven methods referred to as #61 was not listed even in the top 10 methods. Therefore, the final aggregated rank positions for all three OCR engines, together with the relative computation time normalized according to Otsu's thresholding, are presented in Table 4.

Table 1. Comparison of the average F-measure, Levenshtein distance, and Optical Character Recognition (OCR) accuracy values obtained for various binarization methods using the Tesseract OCR engine for 176 document images (three best results shown in bold format).

#	Binarization Method	OCR Measure						Overall Rank
		F-Measure	Rank	Levenshtein Distance	Rank	Accuracy	Rank	
1	Otsu [24]	0.6808	60	1469.88	60	0.5179	60	60
2	Chou [32]	0.8032	57	944.68	58	0.6575	57	57
3	Kittler [22]	0.6173	61	1889.86	61	0.3911	61	61
4	Niblack [8]	0.8838	48	243.39	47	0.7906	48	48
5	Sauvola [49]	0.9428	27	96.79	35	0.8955	27	28
6	Wolf [9]	0.9342	33	142.43	41	0.8800	30	36
7	Bradley (mean) [46]	0.9019	43	245.98	48	0.8217	43	45
8	Bradley (Gaussian) [46]	0.8490	51	557.98	54	0.7319	52	51
9	Feng [50]	0.7438	59	950.16	59	0.5908	59	59
10	Bernsen [44]	0.7673	58	724.68	57	0.6104	58	58
11	Meanthresh	0.8203	55	464.19	52	0.6885	55	54
12	NICK [48]	0.9551	24	43.20	25	0.9144	25	25
13	Wellner [91]	0.9134	40	275.10	50	0.8450	40	42
14	Region (1 layer) [20]	0.8858	46	174.98	42	0.7956	45	44
15	Region (2 layers) [20]	0.9236	38	105.19	36	0.8588	38	38
16	Region (4 layers) [20]	0.9344	31	92.14	31	0.8774	32	30
17	Region (6 layers) [20]	0.9359	30	93.24	32	0.8798	31	29
18	Region (8 layers) [20]	0.9341	34	88.88	29	0.8769	35	33
19	Region (12 layers) [20]	0.9343	32	93.33	33	0.8771	34	34
20	Region (16 layers) [20]	0.9339	35	90.65	30	0.8767	36	35
21	Region (16 layers + MC) [20]	0.9079	42	117.16	39	0.8315	42	41
22	Resampling [19]	0.9557	22	37.13	24	0.9156	23	23
23	Entropy + Otsu [18]	0.8418	53	618.51	56	0.7291	55	54
24	Entropy + Niblack [18]	0.8086	56	491.88	53	0.6758	56	56
25	Entropy + Bradley(Mean) [18]	0.9115	41	94.08	34	0.8405	41	39
26	Entropy + Bradley(Gauss) [18]	0.8908	44	188.71	43	0.8057	44	43
27	Entropy + Meanthresh [18]	0.9404	16	46.93	14	0.8899	17	15
28	SSP [85,86]	0.9402	28	111.99	27	0.8915	29	27
29	Gatos [52]	0.6808	49	1469.88	49	0.5179	49	50
30	Su [67]	0.9332	36	62.21	28	0.9772	33	32
31	Singh [53]	0.8945	25	245.57	23	0.8046	24	24
32	Bataineh [59]	0.3905	52	2578.68	51	0.1860	54	51
33	WAN [56]	0.9504	50	45.39	44	0.9080	50	49
34	ISauvola [54]	0.9459	26	80.53	26	0.8955	26	26
35	OR (#20,#22,#23)	0.9294	37	110.91	37	0.8698	37	37
36	AND (#20,#22,#23)	0.8408	54	615.75	55	0.7337	51	53
37	Voting (#20,#22,#23)	0.9576	18	30.44	17	0.9192	18	19
38	Voting (#5,#12,#22)	0.9585	16	31.35	22	0.9207	16	20
39	Voting (#4,#7,#11)	0.8863	45	236.19	45	0.7950	46	46
40	Voting (#4,#11,#22)	0.8844	47	238.95	46	0.7915	47	47
41	Voting (#7,#20,#23)	0.9206	39	141.19	40	0.8544	39	40
42	Voting (#7,#12,#20,#22,#23)	0.9568	20	29.05	12	0.9177	20	18
43	Voting (#12,#20,#23)	0.9617	8	26.82	8	0.9263	8	7
44	Voting (#12,#22,#27)	0.9586	15	30.88	19	0.9208	15	17
45	Voting (#12,#18,#20,#22,#27)	0.9576	19	31.04	21	0.9188	19	21
46	Voting (#5,#6,#12,#18,#20,#22,#27)	0.9617	7	27.11	9	0.9264	7	6
47	Voting (#16,#22,#23)	0.9556	23	30.93	20	0.9156	22	22
48	Voting (#12,#16,#23)	0.9605	10	27.95	10	0.9243	10	11
49	Voting (#7,#12,#16,#22,#23)	0.9580	17	29.52	13	0.9200	17	16
50	Voting (#20, #23, #34)	**0.9630**	3	26.39	7	**0.9289**	3	3
51	Voting (#20, #27, #31)	0.9602	12	23.31	5	**0.9289**	3	4
52	Voting (#20, #23, #28)	0.9623	6	25.52	6	0.9238	12	7
53	Voting (#22, #27, #34)	0.9597	13	**22.43**	3	0.9277	6	5
54	Voting (#22, #23, #31)	0.9560	21	28.60	11	0.9229	14	15
55	Voting (#22, #27, #28)	0.9630	4	23.11	4	0.9168	21	10
56	Voting (#28, #31, #34)	0.9597	14	30.82	18	0.9232	13	14
57	Voting (#20, #31, #34)	0.9603	11	30.44	16	0.9243	11	13
58	**Voting (#12, #28, #34)**	**0.9660**	1	**20.44**	2	**0.9346**	1	1
59	Voting (#22, #31, #34)	0.9611	9	30.02	15	0.9258	9	12
60	Voting (#4, #7, #28, #31, #34)	0.9626	5	29.88	14	0.9285	5	7
61	Voting (#12, #20, #22, #23, #28, #31, #34)	**0.9653**	2	**18.51**	1	**0.9333**	2	2

Table 2. Comparison of the average F-measure, Levenshtein distance, and OCR accuracy values obtained for various binarization methods using MATLAB's built-in OCR engine for 176 document images (three best results shown in bold format).

#	Binarization Method	OCR Measure						Overall Rank
		F-Measure	Rank	Levenshtein Distance	Rank	Accuracy	Rank	
1	Otsu [24]	0.6306	60	1618.53	60	0.4368	60	60
2	Chou [32]	0.7351	54	1097.47	57	0.5495	55	56
3	Kittler [22]	0.5799	61	2027.23	61	0.3234	61	61
4	Niblack [8]	0.7395	52	455.53	45	0.5787	51	50
5	Sauvola [49]	0.8672	27	267.34	34	0.7655	28	29
6	Wolf [9]	0.8512	30	312.68	39	0.7433	30	31
7	Bradley (mean) [46]	0.8008	41	549.89	49	0.6554	41	43
8	Bradley (Gaussian) [46]	0.7554	47	856.21	55	0.5819	50	51
9	Feng [50]	0.6607	58	1041.55	56	0.4683	57	57
10	Bernsen [44]	0.6640	57	1194.11	58	0.4533	59	58
11	Meanthresh	0.7039	56	663.48	51	0.5212	56	55
12	NICK [48]	0.8589	29	208.87	26	0.7593	29	28
13	Wellner [91]	0.8268	32	470.24	46	0.6985	32	38
14	Region (1 layer) [20]	0.7136	55	455.26	44	0.5515	54	52
15	Region (2 layers) [20]	0.7852	43	286.34	37	0.6499	42	41
16	Region (4 layers) [20]	0.8059	38	249.19	33	0.6790	38	37
17	Region (6 layers) [20]	0.8095	37	249.01	32	0.6841	37	35
18	Region (8 layers) [20]	0.8151	35	239.34	29	0.6919	35	31
19	Region (12 layers) [20]	0.8141	36	241.31	30	0.6908	36	33
20	Region (16 layers) [20]	0.8160	34	242.50	31	0.6932	33	30
21	Region (16 layers + MC) [20]	0.7819	44	305.28	38	0.6429	43	42
22	Resampling [19]	0.8655	28	159.55	18	0.7677	27	26
23	Entropy + Otsu [18]	0.7734	46	786.19	54	0.6161	46	49
24	Entropy + Niblack [18]	0.6372	59	1211.35	59	0.4600	58	59
25	Entropy + Bradley(Mean) [18]	0.8212	33	363.03	41	0.6929	34	36
26	Entropy + Bradley(Gauss) [18]	0.7869	42	525.62	48	0.6398	44	44
27	Entropy + Meanthresh [18]	0.8790	21	149.66	15	0.7879	22	21
28	SSP [85,86]	0.8766	26	235.88	28	0.7802	26	27
29	Gatos [52]	0.7544	48	477.35	47	0.5936	47	46
30	Su [67]	0.8053	39	283.09	35	0.6763	39	39
31	Singh [53]	0.8779	23	185.99	24	0.7822	25	25
32	Bataineh [59]	0.8779	53	185.99	50	0.7822	53	54
33	WAN [56]	0.7461	50	742.53	52	0.5757	52	53
34	ISauvola [54]	0.8835	14	216.48	27	0.7890	20	23
35	OR (#20,#22,#23)	0.8049	40	285.87	36	0.6759	40	40
36	AND (#20,#22,#23)	0.7787	45	765.26	53	0.6269	45	47
37	Voting (#20,#22,#23)	0.8799	18	136.13	9	0.7899	18	14
38	Voting (#5,#12,#22)	0.8767	25	150.70	16	0.7854	24	24
39	Voting (#4,#7,#11)	0.7467	49	437.98	42	0.5887	48	45
40	Voting (#4,#11,#22)	0.7442	51	442.27	43	0.5840	49	47
41	Voting (#7,#20,#23)	0.8310	31	359.44	40	0.7067	31	33
42	Voting (#7,#12,#20,#22,#23)	0.8847	13	134.78	8	0.7977	11	8
43	Voting (#12,#20,#23)	0.8810	17	138.27	11	0.7924	15	12
44	Voting (#12,#22,#27)	0.8792	20	145.94	13	0.7888	21	20
45	Voting (#12,#18,#20,#22,#27)	0.8788	22	130.88	7	0.7892	19	18
46	Voting (#5,#6,#12,#18,#20,#22,#27)	0.8900	8	124.80	4	0.8064	7	5
47	Voting (#16,#22,#23)	0.8798	19	138.04	10	0.7902	17	16
48	Voting (#12,#16,#23)	0.8778	24	139.63	12	0.7868	23	22
49	Voting (#7,#12,#16,#22,#23)	0.8835	15	129.94	6	0.7953	13	10
50	Voting (#20, #23, #34)	0.8966	6	164.73	19	0.8112	6	7
51	Voting (#20, #27, #31)	**0.8993**	2	**118.30**	3	**0.8185**	2	1
52	Voting (#20, #23, #28)	0.8882	10	148.10	14	0.8002	9	9
53	Voting (#22, #27, #34)	0.8966	5	**116.29**	2	0.8141	5	4
54	Voting (#22, #23, #31)	0.8825	16	173.11	20	0.7905	16	19
55	Voting (#22, #27, #28)	**0.8983**	3	**114.48**	1	**0.8178**	3	1
56	Voting (#28, #31, #34)	0.8894	9	189.82	25	0.7991	10	13
57	Voting (#20, #31, #34)	0.8877	11	182.86	22	0.7971	12	14
58	Voting (#12, #28, #34)	0.8982	4	153.56	17	0.8163	4	6
59	Voting (#22, #31, #34)	0.8852	12	181.83	21	0.7932	14	17
60	Voting (#4, #7, #28, #31, #34)	0.8916	7	185.45	23	0.8025	8	11
61	Voting (#12, #20, #22, #23, #28, #31, #34)	**0.9014**	1	129.62	5	**0.8209**	1	1

Table 3. Comparison of the average F-measure, Levenshtein distance, and OCR accuracy values obtained for various binarization methods using GNU Ocrad for 176 document images (three best results shown in bold format).

#	Binarization Method	OCR Measure						Overall Rank
		F-Measure	Rank	Levenshtein Distance	Rank	Accuracy	Rank	
1	Otsu [24]	0.5622	60	2414.45	59	0.2231	60	59
2	Chou [32]	0.6013	56	1884.73	54	0.3316	56	56
3	Kittler [22]	0.5641	59	2487.22	60	0.2019	61	60
4	Niblack [8]	0.6639	43	953.84	36	0.4531	44	42
5	Sauvola [49]	0.7001	35	1136.26	44	0.4938	36	39
6	Wolf [9]	0.7068	32	1009.59	40	0.5083	34	36
7	Bradley (mean) [46]	0.6074	53	1786.23	52	0.3633	54	53
8	Bradley (Gaussian) [46]	0.5745	57	2151.78	58	0.2909	58	58
9	Feng [50]	0.6050	54	1943.19	55	0.3894	52	54
10	Bernsen [44]	0.5020	61	2969.76	61	0.2263	59	61
11	Meanthresh	0.6576	46	1075.15	42	0.4366	47	44
12	NICK [48]	0.7226	22	872.76	28	0.5362	21	21
13	Wellner [91]	0.6796	38	1214.53	46	0.4638	40	43
14	Region (1 layer) [20]	0.6183	51	1057.44	41	0.4038	50	47
15	Region (2 layers) [20]	0.6749	39	800.10	24	0.4780	38	34
16	Region (4 layers) [20]	0.6995	36	735.78	22	0.5098	33	30
17	Region (6 layers) [20]	0.7069	31	727.24	21	0.5198	28	25
18	Region (8 layers) [20]	0.7079	30	721.76	19	0.5210	27	24
19	Region (12 layers) [20]	0.7065	33	724.13	20	0.5195	29	28
20	Region (16 layers) [20]	0.7094	29	720.84	18	0.5237	24	21
21	Region (16 layers + MC) [20]	0.6661	41	824.98	26	0.4641	39	36
22	Resampling [19]	0.7331	18	690.13	16	0.5556	17	17
23	Entropy + Otsu [18]	0.6422	49	1420.09	49	0.4247	49	50
24	Entropy + Niblack [18]	0.6615	45	1962.02	56	0.4586	41	47
25	Entropy + Bradley(Mean) [18]	0.6247	50	1608.03	50	0.3917	51	51
26	Entropy + Bradley(Gauss) [18]	0.6142	52	1699.44	51	0.3740	53	52
27	Entropy + Meanthresh [18]	0.7558	8	573.63	9	0.5889	8	8
28	SSP [85,86]	0.7171	25	884.47	31	0.5225	25	27
29	Gatos [52]	0.6559	47	1178.52	45	0.4373	46	46
30	Su [67]	0.7020	34	814.40	25	0.5122	32	30
31	Singh [53]	0.7180	24	644.81	35	0.5219	26	29
32	Bataineh [59]	0.6041	55	1869.63	53	0.3609	55	55
33	WAN [56]	0.5695	58	2103.19	57	0.3109	57	57
34	ISauvola [54]	0.7109	28	1089.10	43	0.5068	35	36
35	OR (#20,#22,#23)	0.6879	37	883.67	30	0.4897	37	35
36	AND (#20,#22,#23)	0.6493	48	1390.91	48	0.4342	48	49
37	Voting (#20,#22,#23)	0.7565	7	558.05	6	0.5910	6	6
38	Voting (#5,#12,#22)	0.7422	14	675.09	14	0.5683	14	14
39	Voting (#4,#7,#11)	0.6665	40	937.72	33	0.4577	42	39
40	Voting (#4,#11,#22)	0.6648	42	965.40	37	0.4540	43	41
41	Voting (#7,#20,#23)	0.6636	44	1262.76	47	0.4492	45	45
42	Voting (#7,#12,#20,#22,#23)	0.7615	4	552.84	5	0.5980	4	4
43	Voting (#12,#20,#23)	0.7551	9	588.52	11	0.5883	9	10
44	Voting (#12,#22,#27)	0.7419	15	673.88	13	0.5679	15	15
45	Voting (#12,#18,#20,#22,#27)	0.7520	11	584.39	10	0.5846	11	11
46	Voting (#5,#6,#12,#18,#20,#22,#27)	0.7608	5	552.19	4	0.5968	5	5
47	Voting (#16,#22,#23)	0.7529	10	564.89	8	0.5853	10	9
48	Voting (#12,#16,#23)	0.7494	13	595.03	12	0.5799	12	12
49	Voting (#7,#12,#16,#22,#23)	0.7567	6	559.50	7	0.5906	7	7
50	Voting (#20, #23, #34)	0.7316	19	875.44	29	0.5412	19	19
51	Voting (#20, #27, #31)	**0.7673**	2	**530.68**	1	**0.6050**	2	2
52	Voting (#20, #23, #28)	0.7394	16	715.07	17	0.5587	16	16
53	Voting (#22, #27, #34)	**0.7679**	1	**531.49**	2	**0.6061**	1	1
54	Voting (#22, #23, #31)	0.7273	20	841.39	27	0.5386	20	19
55	Voting (#22, #27, #28)	**0.7661**	3	**537.62**	3	**0.6037**	3	3
56	Voting (#28, #31, #34)	0.7159	27	978.32	39	0.5174	31	33
57	Voting (#20, #31, #34)	0.7235	21	935.28	32	0.5287	22	23
58	Voting (#12, #28, #34)	0.7351	17	784.12	23	0.5504	18	18
59	Voting (#22, #31, #34)	0.7218	23	937.92	34	0.5268	23	25
60	Voting (#4, #7, #28, #31, #34)	0.7170	26	973.15	38	0.5189	30	32
61	Voting (#12, #20, #22, #23, #28, #31, #34)	0.7512	12	676.03	15	0.5761	13	13

Table 4. Comparison of the overall rank scores for 3 OCR engines and average computational time relative to Otsu's method obtained for 176 document images.

#	Binarization Method	Final Aggregated Rank	Computation Time (Relative)
1	Otsu [24]	60	1.00
2	Chou [32]	57	5.74
3	Kittler [22]	61	23.30
4	Niblack [8]	46	75.11
5	Sauvola [49]	33	73.73
6	Wolf [9]	36	76.36
7	Bradley (mean) [46]	47	19.62
8	Bradley (Gaussian) [46]	54	241.61
9	Feng [50]	58	215.20
10	Bernsen [44]	59	197.14
11	Meanthresh	51	39.93
12	NICK [48]	25	70.81
13	Wellner [91]	41	187.90
14	Region (1 layer) [20]	49	29.84
15	Region (2 layers) [20]	38	50.23
16	Region (4 layers) [20]	34	92.39
17	Region (6 layers) [20]	31	145.49
18	Region (8 layers) [20]	30	211.87
19	Region (12 layers) [20]	32	325.05
20	Region (16 layers) [20]	29	441.84
21	Region (16 layers + MC) [20]	40	1232.01
22	Resampling [19]	24	12.48
23	Entropy + Otsu [18]	51	664.87
24	Entropy + Niblack [18]	56	755.11
25	Entropy + Bradley(Mean) [18]	42	706.57
26	Entropy + Bradley(Gauss) [18]	45	932.92
27	Entropy + Meanthresh [18]	21	736.67
28	SSP [85,86]	27	4542.24
29	Gatos [52]	48	2413.68
30	Su [67]	35	6016.56
31	Singh [53]	26	59.78
32	Bataineh [59]	54	44.58
33	WAN [56]	53	400.98
34	ISauvola [54]	27	113.69
35	OR (#20,#22,#23)	37	1138.64
36	AND (#20,#22,#23)	50	1134.25
37	Voting (#20,#22,#23)	12	1136.87
38	Voting (#5,#12,#22)	22	159.17
39	Voting (#4,#7,#11)	43	137.30
40	Voting (#4,#11,#22)	44	130.64
41	Voting (#7,#20,#23)	39	1143.63
42	Voting (#7,#12,#20,#22,#23)	9	1224.28
43	Voting (#12,#20,#23)	7	1191.77
44	Voting (#12,#22,#27)	18	817.40
45	Voting (#12,#18,#20,#22,#27)	15	1455.67
46	Voting (#5,#6,#12,#18,#20,#22,#27)	4	1600.90
47	Voting (#16,#22,#23)	14	793.58
48	Voting (#12,#16,#23)	13	858.17
49	Voting (#7,#12,#16,#22,#23)	11	892.77
50	Voting (#20, #23, #34)	7	1249.60
51	Voting (#20, #27, #31)	1	1247.58
52	Voting (#20, #23, #28)	10	5662.15
53	Voting (#22, #27, #34)	2	887.61
54	Voting (#22, #23, #31)	19	801.04
55	Voting (#22, #27, #28)	3	5286.12
56	Voting (#28, #31, #34)	23	4584.69
57	Voting (#20, #31, #34)	15	745.37
58	Voting (#12, #28, #34)	6	4572.60
59	Voting (#22, #31, #34)	20	190.31
60	Voting (#4, #7, #28, #31, #34)	15	4656.10
61	Voting (#12, #20, #22, #23, #28, #31, #34)	4	5880.53

Although not all the results of the tested combinations of various methods are reported in Tables 1–4, it is worth noting that the most successful combinations, leading to the best aggregated rank positions presented in Table 4, contained one of the variants of the multi-layered stack of regions (#20) or the resampling method (#19), as well as an entropy based method (#27). Therefore, the possibilities of the application of these recent approaches in combination with some other algorithms were confirmed. Considering additionally the processing time, a reasonable choice might also be the combination of Methods #22 and #27 with the recent ISauvola algorithm (#34), listed as #53, providing very good results for each of the tested OCR engines in view of Levenshtein distance.

Exemplary results of the binarization of sample documents from the dataset used in experiments are presented in Figures 2–4, where significant differences between some methods may be easily noticed, as well as the relatively high quality of binary images obtained using the proposed approach.

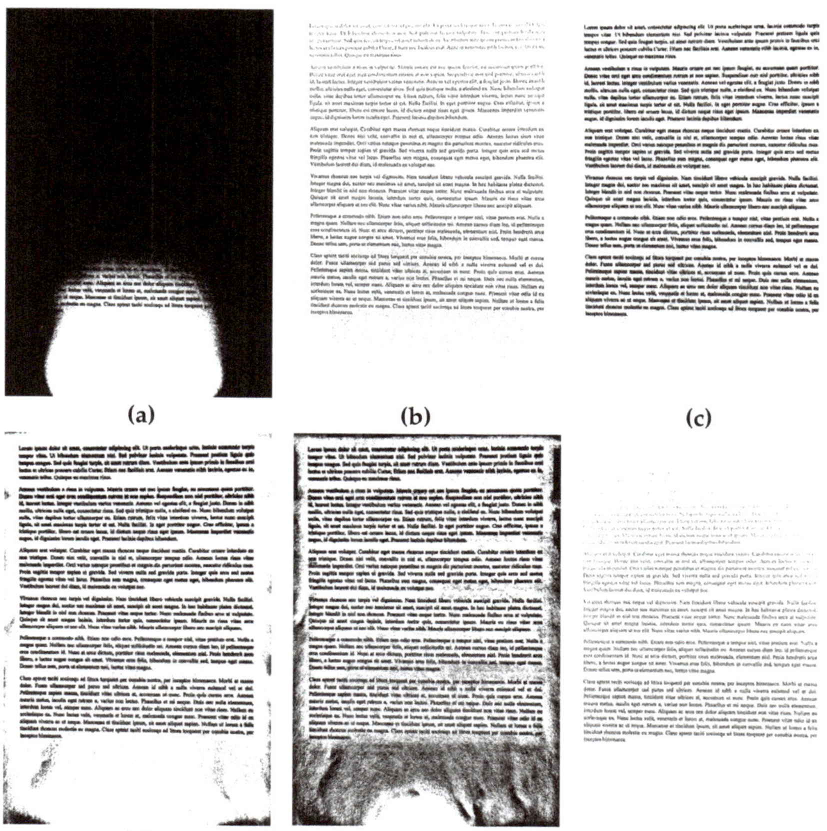

Figure 2. *Cont.*

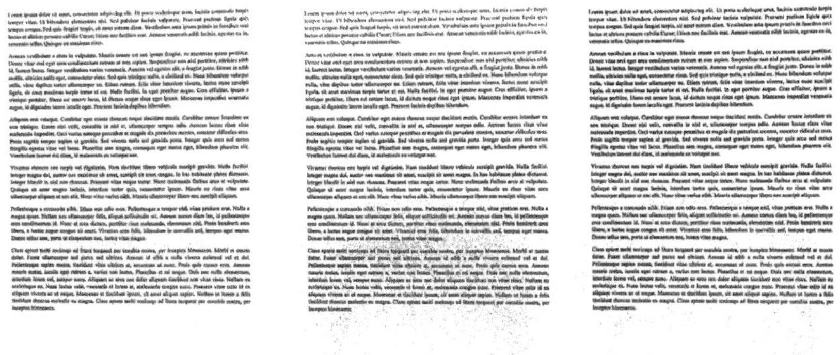

(g) (h) (i)

Figure 2. Binarization results obtained for a sample unevenly illuminated image from the dataset used in the experiments shown in Figure 1a for various methods: (**a**) Otsu, (**b**) Niblack, (**c**) Sauvola, (**d**) Bradley (mean), (**e**) Bernsen, (**f**) Meanthresh, (**g**) NICK , (**h**) stack of regions (16 layers), and (**i**) proposed (#51).

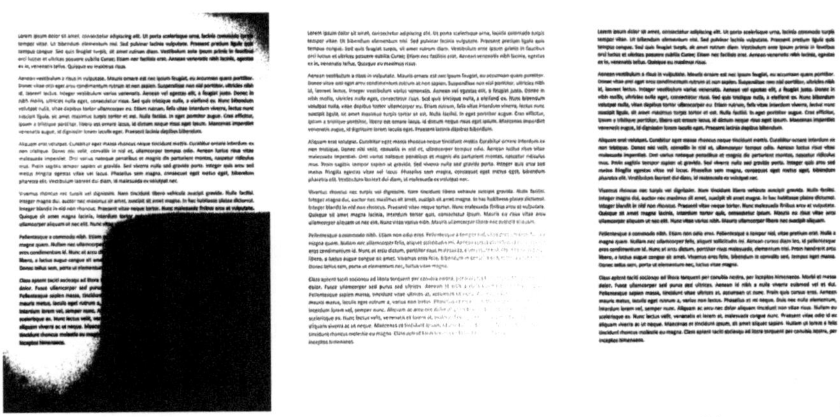

(a) (b) (c)

Figure 3. *Cont.*

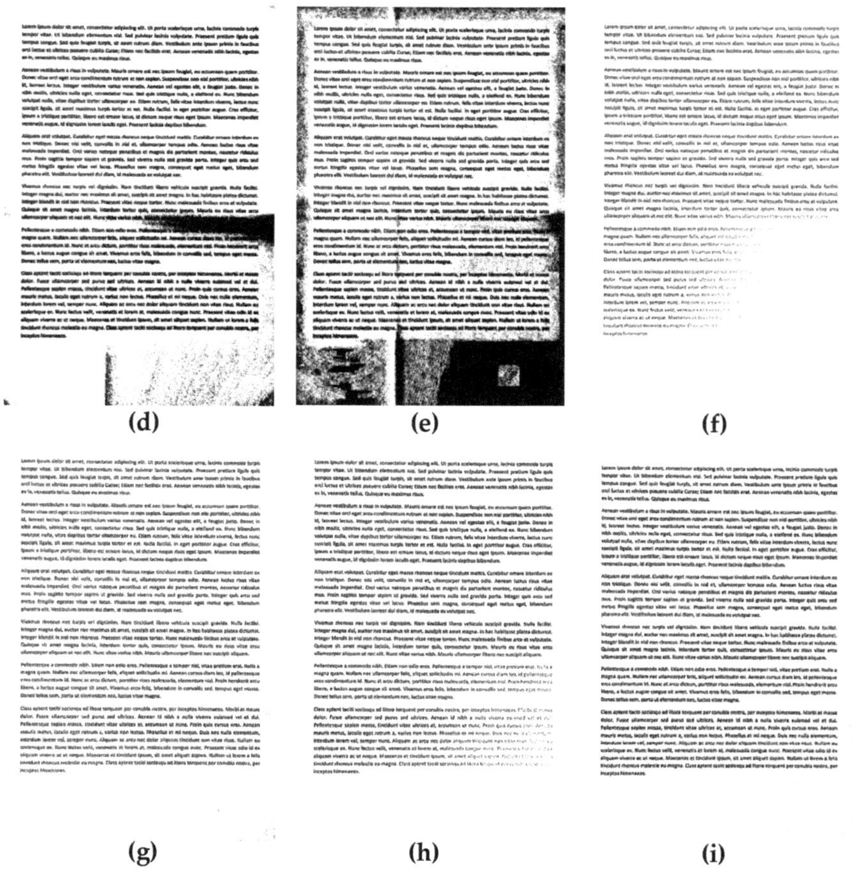

Figure 3. Binarization results obtained for a sample unevenly illuminated image from the dataset used in the experiments shown in Figure 1b for various methods: (**a**) Otsu, (**b**) Niblack, (**c**) Sauvola, (**d**) Bradley (mean), (**e**) Bernsen, (**f**) Meanthresh, (**g**) NICK, (**h**) stack of regions (16 layers), and (**i**) proposed (#51).

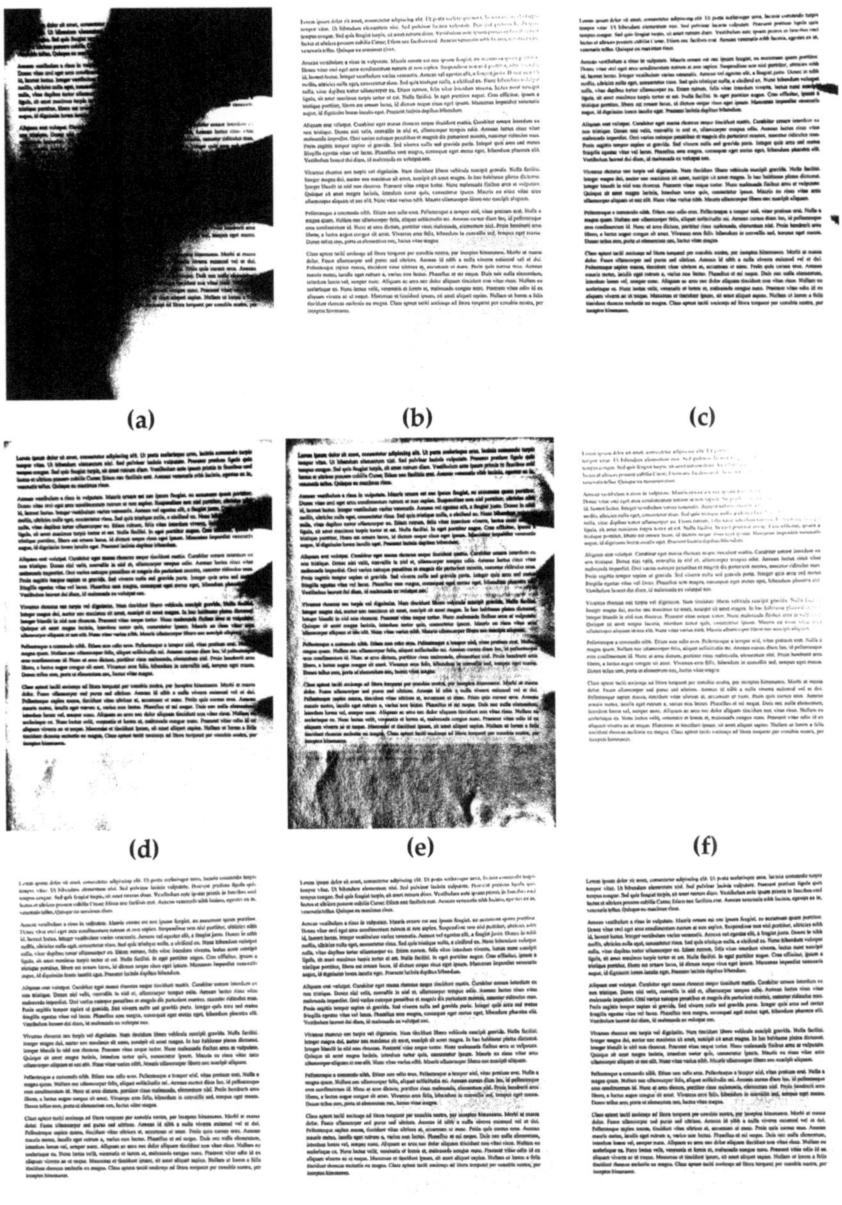

Figure 4. Binarization results obtained for a sample unevenly illuminated image from the dataset used in experiments shown in Figure 1c for various methods: (**a**) Otsu, (**b**) Niblack, (**c**) Sauvola, (**d**) Bradley (mean), (**e**) Bernsen, (**f**) Meanthresh, (**g**) NICK, (**h**) stack of regions (16 layers), and (**i**) proposed (#51).

5. Concluding Remarks

Binarization of non-uniformly illuminated images acquired by camera sensors, especially mounted in mobile devices, in unknown lighting conditions is still a challenging task. Considering the potential applications of the real-time analysis of binary images captured by vision sensors, not only directly related to OCR applications, but also, e.g., to mobile robotics or recognition of the QR codes from natural images, the proposed approach may be an interesting idea providing a reasonable accuracy for various types of illuminations.

The presented experimental results may be extended during future research also by the analysis of the potential applicability of the proposed methods and their combinations for automatic text recognition systems for even more challenging images, e.g., with metallic plates with embossed serial numbers. Another direction for further research may be the investigation of the potential applications of some fuzzy methods [83,84], which may be useful, e.g., for a combination of an even number of algorithms, as well as the use of different weights for each combined method.

Author Contributions: H.M. worked under the supervision of K.O. H.M. prepared the data and sample document images. H.M. and K.O. designed the concept and methodology and proposed the algorithm. H.M. implemented the method, performed the calculations, and prepared the data visualization. K.O. validated the results and wrote the final version of the paper. All authors read and agreed to the published version of the manuscript .

Funding: This research received no external funding.

Acknowledgments: The authors would like to thank the anonymous reviewers for their helpful comments supporting us in improving the current version of the paper and to all researchers who made the codes of their algorithms and the datasets used for their preliminary verification publicly available.

Conflicts of Interest: The authors declare no conflict of interest.

Abbreviations

The following abbreviations are used in this manuscript:

ADAS	Advanced Driver-Assistance System
BHT	Balanced Histogram Thresholding
DIBCO	Document Image Binarization Competition
DSLR	Digital Single Lens Reflex
DRD	Distance Reciprocal Distortion
FAIR	Fast Algorithm for document Image Restoration
GPU	Graphics Processing Unit
GT	Ground Truth
H-DIBCO	Handwritten Document Image Binarization Competition
ICDAR	International Conference on Document Analysis and Recognition
ICFHR	International Conference on Frontiers in Handwriting Recognition
OCR	Optical Character Recognition
QR	Quick Response
SSD	Solid-State Drive
SSP	Structural Symmetric Pixels
SVM	Support Vector Machines

References

1. Okarma, K.; Lech, P. Fast statistical image binarization of color images for the recognition of the QR codes. *Elektron. Ir Elektrotech.* **2015**, *21*, 58–61. [CrossRef]
2. Chen, R.; Yu, Y.; Xu, X.; Wang, L.; Zhao, H.; Tan, H.Z. Adaptive Binarization of QR Code Images for Fast Automatic Sorting in Warehouse Systems. *Sensors* **2019**, *19*, 5466. [CrossRef] [PubMed]
3. Guizzo, E. Superfast Scanner Lets You Digitize Book by Flipping Pages. Available online: https://spectrum.ieee.org/automaton/robotics/robotics-software/book-flipping-scanning (accessed on 20 May 2020).

4. Pratikakis, I.; Zagoris, K.; Karagiannis, X.; Tsochatzidis, L.; Mondal, T.; Marthot-Santaniello, I. ICDAR 2019 Competition on Document Image Binarization (DIBCO 2019). In Proceedings of the 15th IAPR International Conference on Document Analysis and Recognition (ICDAR), Sydney, Australia, 20–25 September 2019; pp. 1547–1556. [CrossRef]
5. Pratikakis, I.; Zagori, K.; Kaddas, P.; Gatos, B. ICFHR 2018 Competition on Handwritten Document Image Binarization (H-DIBCO 2018). In Proceedings of the 16th International Conference on Frontiers in Handwriting Recognition (ICFHR), Niagala Falls, NY, USA, 5–8 August 2018; pp. 489–493. [CrossRef]
6. Chaki, N.; Shaikh, S.H.; Saeed, K. Exploring Image Binarization Techniques. In *Studies in Computational Intelligence*; Springer: New Delhi, India, 2014; Volume 560. [CrossRef]
7. Lins, R.D.; Kavallieratou, E.; Smith, E.B.; Bernardino, R.B.; de Jesus, D.M. ICDAR 2019 Time-Quality Binarization Competition. In Proceedings of the 15th IAPR International Conference on Document Analysis and Recognition (ICDAR), Sydney, Australia, 20–25 September 2019; pp. 1539–1546. [CrossRef]
8. Niblack, W. *An Introduction to Digital Image Processing*; Prentice Hall: Englewood Cliffs, NJ, USA, 1986.
9. Wolf, C.; Jolion, J.M. Extraction and recognition of artificial text in multimedia documents. *Form. Pattern Anal. Appl.* **2004**, *6*, 309–326. [CrossRef]
10. Lins, R.; e Silva, G.P.; Gomes e Silva, A.R. Assessing and Improving the Quality of Document Images Acquired with Portable Digital Cameras. In Proceedings of the 9th International Conference on Document Analysis and Recognition (ICDAR), Parana, Brazil, 23–26 September 2007; Volume 2, pp. 569–573. [CrossRef]
11. Alqudah, M.K.; Bin Nasrudin, M.F.; Bataineh, B.; Alqudah, M.; Alkhatatneh, A. Investigation of binarization techniques for unevenly illuminated document images acquired via handheld cameras. In Proceedings of the International Conference on Computer, Communications, and Control Technology (I4CT), Kuching, Malaysia, 21–23 April 2015; pp. 524–529. [CrossRef]
12. Lins, R.D.; Bernardino, R.B.; de Jesus, D.M.; Oliveira, J.M. Binarizing Document Images Acquired with Portable Cameras. In Proceedings of the 14th IAPR International Conference on Document Analysis and Recognition (ICDAR), Kyoto, Japan, 9–15 November 2017; pp. 45–50. [CrossRef]
13. Pereira, G.; Lins, R.D. PhotoDoc: A Toolbox for Processing Document Images Acquired Using Portable Digital Cameras. In Proceedings of the 2nd International Workshop on Camera-Based Document Analysis and Recognition (CBDAR), Curitiba, Brazil, 22 September 2007; pp. 107–115.
14. Liang, J.; Doermann, D.; Li, H. Camera-based analysis of text and documents: A survey. *Int. J. Doc. Anal. Recognit.* **2005**, *7*, 84–104. [CrossRef]
15. Ntirogiannis, K.; Gatos, B.; Pratikakis, I. Performance evaluation methodology for historical document image binarization. *IEEE Trans. Image Process.* **2013**, *22*, 595–609. [CrossRef]
16. Sokolova, M.; Lapalme, G. A systematic analysis of performance measures for classification tasks. *Inf. Process. Manag.* **2009**, *45*, 427–437. [CrossRef]
17. Lu, H.; Kot, A.; Shi, Y. Distance-reciprocal distortion measure for binary document images. *IEEE Signal Process. Lett.* **2004**, *11*, 228–231. [CrossRef]
18. Michalak, H.; Okarma, K. Improvement of Image Binarization Methods Using Image Preprocessing with Local Entropy Filtering for Alphanumerical Character Recognition Purposes. *Entropy* **2019**, *11*, 286. [CrossRef]
19. Michalak, H.; Okarma, K. Fast Binarization of Unevenly Illuminated Document Images Based on Background Estimation for Optical Character Recognition Purposes. *J. Univ. Comput. Sci.* **2019**, *25*, 627–646.
20. Michalak, H.; Okarma, K. Adaptive Image Binarization Based on Multi-layered Stack of Regions. In *Computer Analysis of Images and Patterns*; Vento, M., Percannella, G., Eds.; Springer International Publishing: Cham, Switzerland, 2019; Volume 11679; pp. 281–293. [CrossRef]
21. dos Anjos, A.; Shahbazkia, H.R. Bi-Level Image Thresholding-A Fast Method. In Proceedings of the 1st International Conference on Biomedical Electronics and Devices (BIOSIGNALS), Funchal, Madeira, Portugal, 28–31 January 2008; pp. 70–76.
22. Kittler, J.; Illingworth, J. Minimum error thresholding. *Pattern Recognit.* **1986**, *19*, 41–47. doi:10.1016/0031-3203(86)90030-0. [CrossRef]
23. Cho, S.; Haralick, R.; Yi, S. Improvement of Kittler and Illingworth's minimum error thresholding. *Pattern Recognit.* **1989**, *22*, 609–617. [CrossRef]
24. Otsu, N. A threshold selection method from gray-level histograms. *IEEE Trans. Syst. Man Cybern.* **1979**, *9*, 62–66. [CrossRef]

25. Kapur, J.; Sahoo, P.; Wong, A. A new method for gray-level picture thresholding using the entropy of the histogram. *Comput. Vis. Gr. Image Process.* **1985**, *29*, 273–285. [CrossRef]
26. Lech, P.; Okarma, K.; Wojnar, D. Binarization of document images using the modified local-global Otsu and Kapur algorithms. *Przegląd Elektrotech.* **2015**, *91*, 71–74. [CrossRef]
27. Ridler, T.; Calvard, S. Picture Thresholding Using an Iterative Selection Method. *IEEE Trans. Syst. Man Cybern.* **1978**, *8*, 630–632. [CrossRef]
28. Xue, J.H.; Zhang, Y.J. Ridler and Calvard's, Kittler and Illingworth's and Otsu's methods for image thresholding. *Pattern Recognit. Lett.* **2012**, *33*, 793–797. [CrossRef]
29. Rosin, P.L. Unimodal thresholding. *Pattern Recognit.* **2001**, *34*, 2083–2096. [CrossRef]
30. Coudray, N.; Buessler, J.L.; Urban, J.P. Robust threshold estimation for images with unimodal histograms. *Pattern Recognit. Lett.* **2010**, *31*, 1010–1019. doi:10.1016/j.patrec.2009.12.025. [CrossRef]
31. Moghaddam, R.F.; Cheriet, M. AdOtsu: An adaptive and parameterless generalization of Otsu's method for document image binarization. *Pattern Recognit.* **2012**, *45*, 2419–2431. [CrossRef]
32. Chou, C.H.; Lin, W.H.; Chang, F. A binarization method with learning-built rules for document images produced by cameras. *Pattern Recognit.* **2010**, *43*, 1518–1530. [CrossRef]
33. Xiong, W.; Xu, J.; Xiong, Z.; Wang, J.; Liu, M. Degraded historical document image binarization using local features and support vector machine (SVM). *Optik* **2018**, *164*, 218–223. [CrossRef]
34. Michalak, H.; Okarma, K. Region based adaptive binarization for optical character recognition purposes. In Proceedings of the International Interdisciplinary PhD Workshop (IIPhDW), Świnoujście, Poland, 9–12 May 2018; pp. 361–366. [CrossRef]
35. Michalak, H.; Okarma, K. Fast adaptive image binarization using the region based approach. In *Artificial Intelligence and Algorithms in Intelligent Systems*; Silhavy, R., Ed.; Springer: New York, NY, USA, 2019; Volume 764, pp. 79–90. [CrossRef]
36. Pun, T. A new method for grey-level picture thresholding using the entropy of the histogram. *Signal Process.* **1980**, *2*, 223–237. [CrossRef]
37. Pun, T. Entropic thresholding, a new approach. *Comput. Gr. Image Process.* **1981**, *16*, 210–239. [CrossRef]
38. Tian, X.; Hou, X. A Tsallis-entropy image thresholding method based on two-dimensional histogram obique segmentation. In Proceedings of the 2009 WASE International Conference on Information Engineering, Taiyuan, Chanxi, China, 10–11 July 2009; Volume 1, pp. 164–168. [CrossRef]
39. Le, T.H.N.; Bui, T.D.; Suen, C.Y. Ternary entropy-based binarization of degraded document images using morphological operators. In Proceedings of the 11th IAPR International Conference on Document Analysis and Recognition (ICDAR), Beijing, China, 18–21 September 2011; pp. 114–118. [CrossRef]
40. Fan, J.; Wang, R.; Zhang, L.; Xing, D.; Gan, F. Image sequence segmentation based on 2D temporal entropic thresholding. *Pattern Recognit. Lett.* **1996**, *17*, 1101–1107. [CrossRef]
41. Abutaleb, A.S. Automatic thresholding of gray-level pictures using two-dimensional entropy. *Comput. Vis. Gr. Image Process.* **1989**, *47*, 22–32. [CrossRef]
42. Tang, K.; Yuan, X.; Sun, T.; Yang, J.; Gao, S. An improved scheme for minimum cross entropy threshold selection based on genetic algorithm. *Knowl.-Based Syst.* **2011**, *24*, 1131–1138. [CrossRef]
43. Li, J.; Tang, W.; Wang, J.; Zhang, X. A multilevel color image thresholding scheme based on minimum cross entropy and alternating direction method of multipliers. *Optik* **2019**, *183*, 30–37. [CrossRef]
44. Bernsen, J. Dynamic thresholding of grey-level images. In Proceedings of the 8th International Conference on Pattern Recognition (ICPR), Paris, France, 27–31 October 1986; pp. 1251–1255.
45. Yang, L.; Feng, Q. The Improvement of Bernsen Binarization Algorithm for QR Code Image. In Proceedings of the 5th International Conference on Cloud Computing and Intelligence Systems (CCIS), Nanjing, China, 23–25 November 2018; pp. 931–934. [CrossRef]
46. Bradley, D.; Roth, G. Adaptive thresholding using the integral image. *J. Gr. Tools* **2007**, *12*, 13–21. [CrossRef]
47. Shafait, F.; Keysers, D.; Breuel, T.M. Efficient implementation of local adaptive thresholding techniques using integral images. In Proceedings of the Document Recognition and Retrieval XV, San Jose, CA, USA, 27–31 January 2008; Volume 6815, [CrossRef]
48. Khurshid, K.; Siddiqi, I.; Faure, C.; Vincent, N. Comparison of Niblack inspired binarization methods for ancient documents. In *Document Recognition and Retrieval XVI*; SPIE: Bellingham, WA, USA, 2009; Volume 7247, pp. 7247–7249. [CrossRef]

49. Sauvola, J.; Pietikäinen, M. Adaptive document image binarization. *Pattern Recognit.* **2000**, *33*, 225–236. [CrossRef]
50. Feng, M.L.; Tan, Y.P. Adaptive binarization method for document image analysis. In Proceedings of the 2004 IEEE International Conference on Multimedia and Expo (ICME), Taipei, Taiwan, 27–30 June 2004; Volume 1, pp. 339–342. [CrossRef]
51. Lazzara, G.; Géraud, T. Efficient multiscale Sauvola's binarization. *Int. J. Doc. Anal. Recognit.* **2014**, *17*, 105–123. [CrossRef]
52. Gatos, B.; Pratikakis, I.; Perantonis, S. Adaptive degraded document image binarization. *Pattern Recognit.* **2006**, *39*, 317–327. [CrossRef]
53. Singh, T.R.; Roy, S.; Singh, O.I.; Sinam, T.; Singh, K.M. A New Local Adaptive Thresholding Technique in Binarization. *IJCSI Int. J. Comput. Sci. Issues* **2011**, *8*, 271–277.
54. Hadjadj, Z.; Meziane, A.; Cherfa, Y.; Cheriet, M.; Setitra, I. ISauvola: Improved Sauvola's Algorithm for Document Image Binarization. In *Image Analysis and Recognition*; Campilho, A., Karray, F., Eds.; Springer International Publishing: Cham, Switzerland, 2016; Volume 9730, pp. 737–745. [CrossRef]
55. He, Y.; Yang, Y. An Improved Sauvola Approach on QR Code Image Binarization. In Proceedings of the 11th International Conference on Advanced Infocomm Technology (ICAIT), Jinan, China, 18–20 October 2019; pp. 6–10. [CrossRef]
56. Azani Mustafa, W.; Kader, M.M.M.A. Binarization of Document Image Using Optimum Threshold Modification. *J. Phys. Conf. Ser.* **2018**, *1019*, 012022. [CrossRef]
57. Kulyukin, V.; Kutiyanawala, A.; Zaman, T. Eyes-free barcode detection on smartphones with Niblack's binarization and Support Vector Machines. In Proceedings of the 16th International Conference on Image Processing, Computer Vision, and Pattern Recognition (IPCV'2012), Las Vegas, NV, USA, 16–19 July 2012; Volume 1, pp. 284–290.
58. Samorodova, O.A.; Samorodov, A.V. Fast implementation of the Niblack binarization algorithm for microscope image segmentation. *Pattern Recognit. Image Anal.* **2016**, *26*, 548–551. [CrossRef]
59. Bataineh, B.; Abdullah, S.N.H.S.; Omar, K. An adaptive local binarization method for document images based on a novel thresholding method and dynamic windows. *Pattern Recognit. Lett.* **2011**, *32*, 1805–1813. [CrossRef]
60. Mysore, S.; Gupta, M.K.; Belhe, S. Complex and degraded color document image binarization. In Proceedings of the 3rd International Conference on Signal Processing and Integrated Networks (SPIN), Noida, India, 11–12 February 2016; pp. 157–162. [CrossRef]
61. Leedham, G.; Yan, C.; Takru, K.; Tan, J.H.N.; Mian, L. Comparison of some thresholding algorithms for text/background segmentation in difficult document images. In Proceedings of the 7th International Conference on Document Analysis and Recognition (ICDAR), Edinburgh, UK, 6 August 2003; pp. 859–864. [CrossRef]
62. Sezgin, M.; Sankur, B. Survey over image thresholding techniques and quantitative performance evaluation. *J. Electron. Imaging* **2004**, *13*, 146–165. [CrossRef]
63. Shrivastava, A.; Srivastava, D.K. A review on pixel-based binarization of gray images. In *ICICT 2015*; Springer: Singapore, 2016; Volume 439, pp. 357–364. [CrossRef]
64. Saxena, L.P. Niblack's binarization method and its modifications to real-time applications: A review. *Artif. Intell. Rev.* **2017**, 1–33. [CrossRef]
65. Mustafa, W.A.; Kader, M.M.M.A. Binarization of document images: A comprehensive review. *J. Phys. Conf. Series* **2018**, *1019*, 012023, [CrossRef]
66. Sulaiman, A.; Omar, K.; Nasrudin, M.F. Degraded historical document binarization: a review on issues, challenges, techniques, and future directions. *J. Imaging* **2019**, *5*, 48. [CrossRef]
67. Su, B.; Lu, S.; Tan, C.L. Robust document image binarization technique for degraded document images. *IEEE Trans. Image Process.* **2013**, *22*, 1408–1417. [CrossRef]
68. Bag, S.; Bhowmick, P. Adaptive-interpolative binarization with stroke preservation for restoration of faint characters in degraded documents. *J. Vis. Commun. Image Represent.* **2015**, *31*, 266–281. [CrossRef]
69. Howe, N.R. A Laplacian energy for document binarization. In Proceedings of the 11th IAPR International Conference on Document Analysis and Recognition (ICDAR), Beijing, China, 18–21 September 2011; pp. 6–10. [CrossRef]

70. Lu, S.; Su, B.; Tan, C.L. Document image binarization using background estimation and stroke edges. *Int. J. Doc. Anal. Recognit.* **2010**, *13*, 303–314. [CrossRef]
71. Erol, B.; Antúnez, E.R.; Hull, J.J. HOTPAPER: multimedia interaction with paper using mobile phones. In Proceedings of the 16th International Conference on Multimedia 2008, Vancouver, BC, Canada, 26–31 October 2008; pp. 399–408. [CrossRef]
72. Okamoto, A.; Yoshida, H.; Tanaka, N. A binarization method for degraded document images with morphological operations. In Proceedings of the 13th IAPR International Conference on Machine Vision Applications (MVA), Kyoto, Japan, 20–23 May 2013; pp. 294–297.
73. Khitas, M.; Ziet, L.; Bouguezel, S. Improved degraded document image binarization using median filter for background estimation. *Elektron. Ir Elektrotech.* **2018**, *24*, 82–87. doi:10.5755/j01.eie.24.3.20982. [CrossRef]
74. Wen, J.; Li, S.; Sun, J. A new binarization method for non-uniform illuminated document images. *Pattern Recognit.* **2013**, *46*, 1670–1690. [CrossRef]
75. Mitianoudis, N.; Papamarkos, N. Document image binarization using local features and Gaussian mixture modeling. *Image Vis. Comput.* **2015**, *38*, 33–51. [CrossRef]
76. Chen, Y.; Wang, L. Broken and degraded document images binarization. *Neurocomputing* **2017**, *237*, 272–280. [CrossRef]
77. Lelore, T.; Bouchara, F. Super-resolved binarization of text based on the FAIR algorithm. In Proceedings of the 11th IAPR International Conference on Document Analysis and Recognition (ICDAR), Beijing, China, 18–21 September 2011; pp. 839–843. [CrossRef]
78. Yazid, H.; Arof, H. Gradient based adaptive thresholding. *J. Vis. Commun. Image Represent.* **2013**, *24*, 926–936. [CrossRef]
79. Feng, S. A novel variational model for noise robust document image binarization. *Neurocomputing* **2019**, *325*, 288–302. [CrossRef]
80. Almeida, M.; Lins, R.D.; Bernardino, R.; Jesus, D.; Lima, B. A New Binarization Algorithm for Historical Documents. *J. Imaging* **2018**, *4*, 27. [CrossRef]
81. Tensmeyer, C.; Martinez, T. Document image binarization with fully convolutional neural networks. In Proceedings of the 14th IAPR International Conference on Document Analysis and Recognition (ICDAR), Kyoto, Japan, 9–15 November 2017; pp. 99–104. [CrossRef]
82. Vo, Q.N.; Kim, S.H.; Yang, H.J.; Lee, G. Binarization of degraded document images based on hierarchical deep supervised network. *Pattern Recognit.* **2018**, *74*, 568–586. [CrossRef]
83. Bogiatzis, A.; Papadopoulos, B. Producing fuzzy inclusion and entropy measures and their application on global image thresholding. *Evol. Syst.* **2018**, *9*, 331–353. [CrossRef]
84. Bogiatzis, A.; Papadopoulos, B. Global Image Thresholding Adaptive Neuro-Fuzzy Inference System Trained with Fuzzy Inclusion and Entropy Measures. *Symmetry* **2019**, *11*, 286. [CrossRef]
85. Jia, F.; Shi, C.; He, K.; Wang, C.; Xiao, B. Document Image Binarization Using Structural Symmetry of Strokes. In Proceedings of the 15th International Conference on Frontiers in Handwriting Recognition (ICFHR), Shenzhen, China, 23–26 October 2016; pp. 411–416. [CrossRef]
86. Jia, F.; Shi, C.; He, K.; Wang, C.; Xiao, B. Degraded document image binarization using structural symmetry of strokes. *Pattern Recognit.* **2018**, *74*, 225–240. [CrossRef]
87. Lins, R.D.; Bernardino, R.B.; de Jesus:, D.M. A Quality and Time Assessment of Binarization Algorithms. In Proceedings of the 15th IAPR International Conference on Document Analysis and Recognition (ICDAR), Sydney, Australia, 20–25 September 2019; pp. 1444–1450. [CrossRef]
88. Yoon, Y.; Ban, K.D.; Yoon, H.; Lee, J.; Kim, J. Best combination of binarization methods for license plate character segmentation. *ETRI J.* **2013**, *35*, 491–500. [CrossRef]
89. Smith, R. An Overview of the Tesseract OCR Engine. In Proceedings of the 9th International Conference on Document Analysis and Recognition (ICDAR), Parana, Brazil, 23–26 September 2007; Volume 2, pp. 629–633. [CrossRef]

90. Deng, F.; Wu, Z.; Lu, Z.; Brown, M.S. Binarizationshop: A user assisted software suite for converting old documents to black-and-white. In Proceedings of the Annual Joint Conference on Digital Libraries, Gold Coast, Queensland, Australia, 21–25 June 2010; pp. 255–258. [CrossRef]
91. Wellner, P.D. *Adaptive Thresholding for the DigitalDesk;* Technical Report EPC 1993-110; Rank Xerox Ltd.: Cambridge, UK, July 1993.

sensors

MDPI

Communication

Converting a Common Low-Cost Document Scanner into a Multispectral Scanner

Zohaib Khan [1,†], Faisal Shafait [2,3,†] and Ajmal Mian [4,*]

1 School of Information Technology and Mathematical Sciences, University of South Australia, Adelaide, SA 5095, Australia
2 School of Electrical Engineering and Computer Science (SEECS), National University of Science and Technology (NUST), Islamabad 44000, Pakistan
3 Deep Learning Laboratory, National Center of Artificial Intelligence, Islamabad 44000, Pakistan
4 Department of Computer Science and Software Engineering, The University of Western Australia, Crawley, WA 6009, Australia
* Correspondence: ajmal.mian@uwa.edu.au; Tel.: +61-8-6488-2702
† Previous address: Department of Computer Science and Software Engineering, The University of Western Australia, Crawley, WA 6009, Australia.

Received: 20 May 2019; Accepted: 18 July 2019; Published: 20 July 2019

Abstract: Forged documents and counterfeit currency can be better detected with multispectral imaging in multiple color channels instead of the usual red, green and blue. However, multispectral cameras/scanners are expensive. We propose the construction of a low cost scanner designed to capture multispectral images of documents. A standard sheet-feed scanner was modified by disconnecting its internal light source and connecting an external multispectral light source comprising of narrow band light emitting diodes (LED). A document was scanned by illuminating the scanner light guide successively with different LEDs and capturing a scan of the document. The system costs less than a hundred dollars and is portable. It can potentially be used for applications in verification of questioned documents, checks, receipts and bank notes.

Keywords: multispectral imaging; document scanning; portable sensor

1. Introduction

Forensic analysis of questioned documents involves a broad range of activities [1]. This includes establishing whether a document originated from a particular source, is backdated, forged or willfully manipulated. Disputes resolution over the authenticity of bank checks [2], purchase receipts, currency notes [3] or seals in agreements [4] can involve overwhelmingly complex legal procedures. In other cases, verification of the genuineness of the document source (written or printed) is also of significant importance to fraud detection [5]. The estimated age of a testament (will) can sometimes play a crucial role in the resolution of inheritance claims [6].

Traditionally, forensic scientists make empirical or experimental observations about a suspicious portion of the document in a forensic laboratory. The observations are then coupled with expert opinions to be presentable in a court-of-law. As this process largely relies on individual expertise and analysis, its consequences may be critical to the rights of a person, business or an organization. There is an interest in mechanisms for pre-examination of questioned documents before legally pursuing and bearing substantial costs in a court-of-law. Computerized forensic analysis has recently paved the way for automatic document forgery detection using *multispectral imaging* [7,8]. Multispectral or hyperspectral document scanners are generally comprised of bulky apparatus and require specialized laboratory environment for operation. This opens the need for the development of a portable multispectral document scanning system.

There are different ways of capturing multispectral images of a scene [9]. The most suitable method can depend on the target application. A *spatial scanner* simultaneously captures (x, λ) dimensions of a scene, whereas the y dimension is captured by the movement of the sensor or the scene. It is suitable for scenarios where either the scene or the sensing platform is moving such as in remote sensing. A flatbed multispectral document scanner can be regarded as a kind of spatial scanner, an example of which exists as a commercial device [10]. Flatbed scanners have a compact construction, however their scanning area is generally limited to an A4 size paper. Benchtop hyperspectral scanners have a similar operational procedure, and capture a relatively larger number of channels. The flexibility of a bench-top construction allows documents as well as other non-planar objects of interest to be scanned by the same device, at the expense of longer times per scan. Benchtop hyperspectral scanners have been shown to be useful for visual enhancement of old documents where a non-contact scanning mechanism may be preferred [11].

A *spectral scanner* simultaneously captures (x, y) dimensions of a scene, whereas the λ dimension is captured by spectral tuning [12]. It is specifically useful in a setup where both the scene and the sensing platform are stationary. The most common construction of a spectral scanner comprises a monochrome camera with a chromatic filter. A filter may be mechanically interchangeable using a wheel, which can be slow and require manual intervention. Such a device has been used for historical document image restoration [13]. A filter can also be tunable, thereby providing faster image scanning. The use of an accousto-optic tunable filter has been demonstrated for the purpose of document authentication [14] and liquid crystal tunable filter has shown to be effective in analysis of inks in documents [15]. However, camera captured documents suffer from image distortion due to perspective view, as well as non-uniformity of illumination. Moreover, the effective spatial resolution of a camera based multispectral document capture system may be much lower than a conventional document scanner.

In contrast, a *snapshot spatio-spectral sensor* simultaneously captures both spatial and spectral (x, y, λ) dimensions of a scene eliminating the need for scanning [16]. This method can effectively be used in conditions where the scene and the sensing platform are simultaneously moving. However, its complex sensor design incurs heavy costs limiting its use in applications such as in-vivo imaging of organisms [17].

Previously, we proposed a spectral scanning system for capturing multispectral images of a document [18]. Despite the simplicity of a static scene and the sensor, the system was prone to artifacts of camera captured imaging (illumination, perspective etc.) [19]. In this work, we propose a *spatio-spectral scanner* for capturing multispectral image of a document using a sheet-feed scanner, thus avoiding the problems associated with cameras. It captures one spatial dimension x, whereas the (y, λ) dimensions are sequentially acquired by feeding the document and tuning illumination spectrum, respectively. In the following section, we describe the proposed multispectral document scanning system, in terms of its electrical, spectral and optical design.

2. Materials and Methods

The main components of the proposed multispectral document scanner are an external multispectral light source and a standard document scanner.

2.1. Multispectral Light Source

A broadband source of light (e.g., incandescent or fluorescent) reflects the average response of a scene over a wide spectral range, and therefore achieves a low spectral fidelity. A multispectral source produces light in narrow spectral bands, attaining a high spectral fidelity. Light Emitting Diodes (LEDs) can provide such selectivity required in the spectral profile of a multispectral light source. Another favorable characteristic of LEDs is that they are highly energy-efficient compared to other sources of light.

2.1.1. Electrical Design

The electrical schematic of the multispectral light source is given in Figure 1. It consists of a constant current source (i1) connected to narrow-band LEDs (d1–d7) via switches (s1–s7). The constant current source limits the current from surpassing the absolute maximum current rating of the LED. It also makes an LED glow with the same luminous power and spectral profile, making the system reliable. However, an inadvertent connection of multiple switches simultaneously can result in the current being divided into several LEDs.

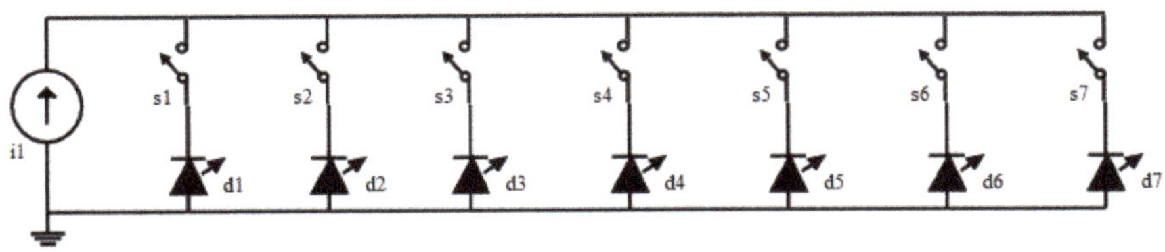

Figure 1. A schematic diagram of the multispectral light source showing the connection layout of the LEDs (d1–d7) and the constant current source (i1) via switches (s1–s7).

To ensure only one LED is powered at a time, a unipolar multi-way rotary switch is included in the design. It provides non-shorting, break-before-make contacts, to avoid overloading of the source with multiple LEDs during switching. It can handle high currents of up to 500 mA @ 250 V ac/dc. The switch and its terminal positions as viewed from the knob end of the spindle are shown in Figure 2. Terminal A (middle) is connected to the positive end of the constant current source. Terminals 1–7 are connected to the positive terminals of d1–d7, respectively. If more spectral bands are desired to be captured, the corresponding LEDs can be conveniently connected to Terminals 8–12, which are currently not utilized.

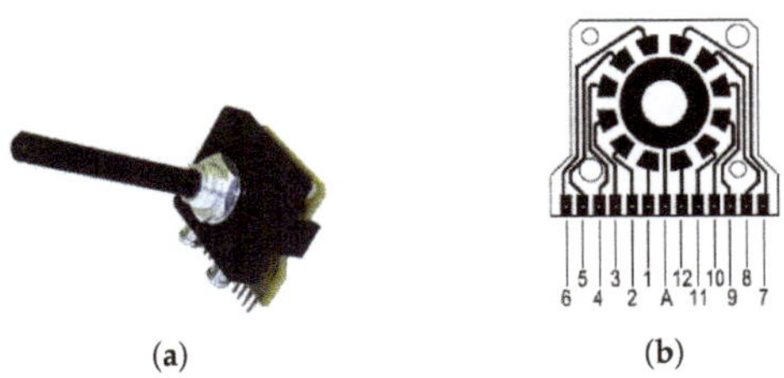

Figure 2. Schematic and assembled unit of a 30 degree indexing, 12 way unipolar switch: (**a**) *PT-6015* rotary switch from *Lorlin Electronics Ltd.*; Sussex, England and (**b**) schematic diagram of connection terminals.

Two constant current sources were designed depending on availability of a low or high input voltage source. The electrical schematic of the sources and their assembled form are shown in Figure 3.

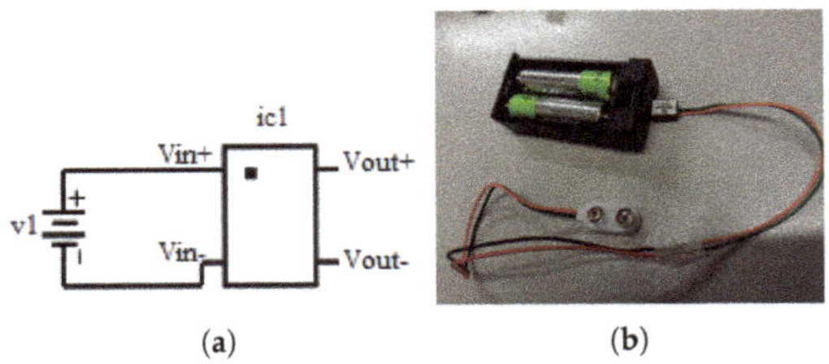

Figure 3. *Cont.*

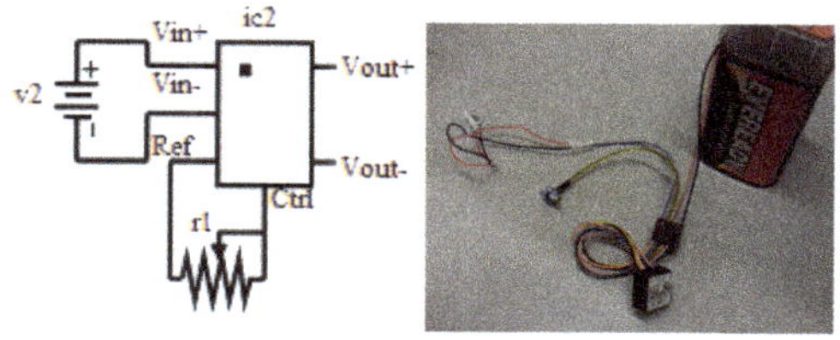

(c) (d)

Figure 3. Schematic and assembly of constant current sources: (**a**) input terminals (V_{in+}, V_{in-}) of the MicroPuck driver (ic1) are connected to a low voltage source (v1 = 0.8–3 Vdc); (**b**) assembled low input voltage–constant current source; (**c**) input terminals (V_{in+}, V_{in-}) of the BuckPuck driver (ic2) are connected to a high voltage source (v1 = 7–32 Vdc) and the potentiometer (r1) allows dimming control (Ctrl) via internal reference (Ref); and (**d**) assembled high input voltage–constant current source.

The low input voltage–constant current source uses a MicroPuck LED Power Module which can provide a constant (350 mA) current to a single LED. The driver has two input pins (V_{in+}, V_{in-}) and two output pins (V_{out+}, V_{out-}). The miniature design allows use of one or two AA sized batteries to power the module. It provides the maximum current to the LEDs while mimicking the light drop-off of an incandescent bulb, which dims as the batteries drain. However, the current drops only at very low voltages, allowing maximum operational time.

The high input voltage–constant current source uses a BuckPuck LED Power Module which can provide a constant (350 mA) current to multiple LEDs. The module has four input pins (V_{in+}, V_{in-}, Ref, Ctrl) and two output pins (V_{out+}, V_{out-}). The module provides manual dimming control through a potentiometer which uses internal reference from the BuckPuck driver. It also has built-in protection for open-circuit and short-circuit.

2.1.2. Spectral Profile

The choice of colored LEDs is important for description of the spectral profile of the multispectral light source. The spectral characteristics are characterized by two main parameters, i.e., the center wavelength and the spectral bandwidth. The relative spectral power distribution of the LEDs is given in Figure 4. These LEDs cover the majority of the range of visible electromagnetic spectrum (400–700 nm) at approximately regular intervals. The spectral parameters of the LEDs are provided in Table 1. Note that the LEDs are spread across the spectrum with sufficiently narrow-bands and high luminous power, which makes an effective multispectral light source.

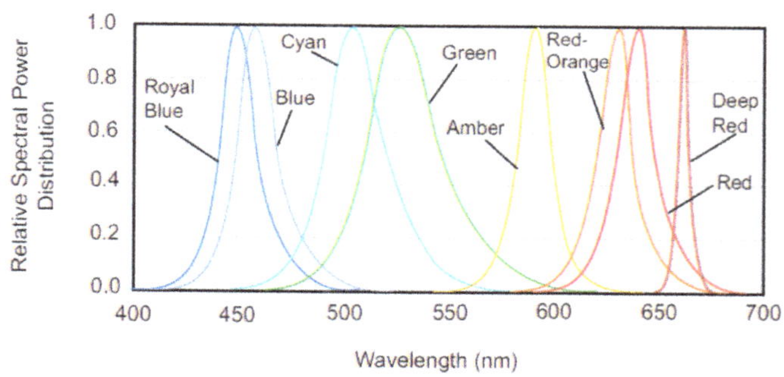

Figure 4. The relative spectral power distribution of the *Philips Luxeon Rebel* LEDs used in this study.

Table 1. Specifications of *Luxeon Rebel* series LEDs at 350 mA.

Color	Center Wavelength (nm)	Bandwidth (nm)	Flux or Power (lm,mW)	Part Number Model
Deep Red	655	20	360	LXM3-PD01
Red	627	20	48	LXM2-PD01-0040
Red-Orange	617	20	56	LXML-PH01-0050
Amber	590	20	48	LXML-PL01-0040
Green	530	30	95	LXML-PM01-0090
Cyan	505	30	76	LXML-PE01-0070
Blue	470	20	41	LXML-PB01-0040
Royal Blue	447.5	20	1030 ‡	LXML-PR02-A900
Neutral White	-	-	180	LXML-PWN1-0100

‡ tested at 700 mA.

Although the range of selected LEDs is in the visible spectrum, the proposed scanner design is generic and not restricted to the visible spectral range. Extension of the spectral range is a matter of adding LEDs (e.g., UV or infrared) in the proposed multispectral light source.

2.1.3. Optical Configuration

The purpose of optical assembly is to transmit multispectral light into a flexible light guide, connected to the scanner light guide. Concentration optics are suitable for beam insertion into fiber optic bundles or light guides. Two different optical arrangements were proposed for multispectral light source, as shown in Figure 5. The two optical configurations after the assembly are shown in Figure 6.

In the linear arrangement, an LED is pre-soldered to a base with anode(+) and cathode(−) connections at the locations shown in Figure 5a. A fiber beam lens (*Carclo Optics, Aylesbury, England*), shown in Figure 5b, focuses light from LED into an eight-degree narrow beam at a focal distance of 11 mm. The diameter of the lens is 20 mm and conforms to the LED base. It requires a circular lens holder, as shown in Figure 5c, which is affixed to the base using a double-sided tape. The holder positions the lens at an appropriate distance from the LED to obtain the maximum luminous transmission. Multiple such units, each with a different colored LED, together make a multispectral light source.

In the array arrangement, multiple LEDs are pre-soldered to a single base with separate anode(+) and cathode(−) connections for each unit at the locations shown in Figure 5d. A cluster concentrator optic (*Polymer Optics Ltd.*), Berkshire, England shown in Figure 5e, focuses light from seven LEDs into a 12-mm narrow beam at a focal distance of 25 mm. It is made of an optical grade poly-carbonate material for thermal stability and system durability, which results in a high light collection efficiency (85%). The use of the array LEDs and the cluster optic makes the light source compact and rigid.

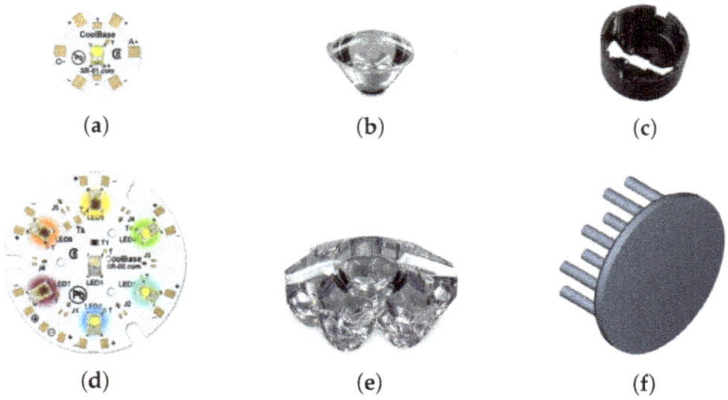

Figure 5. Components of the two optical configurations of multispectral light source: (**a**) single LED assembly; (**b**) fiber coupler concentrator lens; (**c**) circular lens holder; (**d**) seven-LED array assembly; (**e**) multi-cell cluster concentrator optic; and (**f**) natural convection heat sink.

The use of high-power LED array can introduce significant overheating if it is not correctly catered for. A heat sink is an affordable device for maintaining near constant temperature of LEDs for long periods of operation. The *CN40-15B* heat sink from *ALPHA Co. Ltd.*, Shizuoka, Japan has a 40-mm round base with 15-mm legs, as shown in Figure 5f. It has the highest thermal efficiency in the *CN40* series of heat sinks.

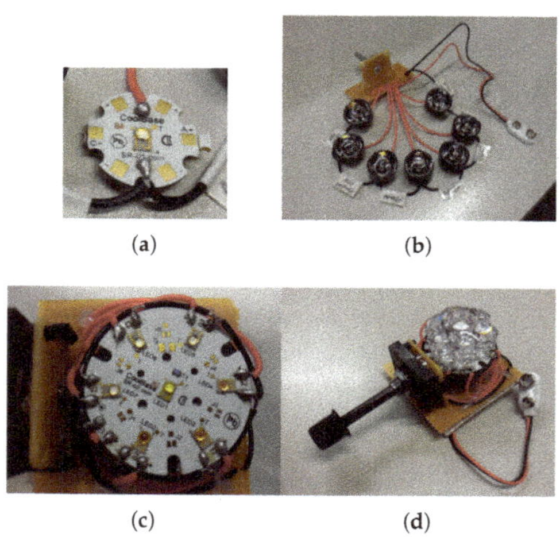

Figure 6. The assembly of linear and array optical configurations: (**a**) single LED base with connections to switch; (**b**) patched assemblies with fiber beam lens in linear configuration; (**c**) array LED base with connections to switch; and (**d**) patched assembly with cluster concentrator lens in array configuration.

2.2. Document Scanner

Connection of an external source of light can be intrusive to the movement of the scanner carriage unit in a flatbed scanner, which may cause discrepancies in the scanned image. In contrast, a sheet-feed scanner allows integration with an external source of light without being intrusive to the scanner operation. Since the scanning unit of a sheet-feed scanner is stationary, its operation is not affected by connection to an external source of light. Moreover, the size of a sheet-feed scanner is mainly determined by the shorter edge of the supported page size, which makes it compact and portable, as shown in Figure 7a. A sheet-feed mechanism is therefore preferred over a flatbed construction to form the basis of a multispectral document scanner.

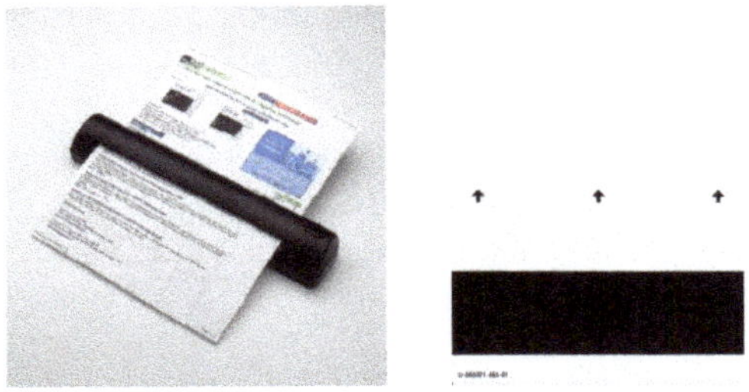

(a) (b)

Figure 7. An automatic feed portable document scanner can be converted into a multi-spectral document scanner: (**a**) the *DSmobile600* sheet-feed portable document scanner from *Brother Mobile Solutions Inc.*, Westminister, CO, USA used in this work; and (**b**) a standard black and white reference sheet for calibration. Arrows indicate the direction of feeding into the scanner.

2.2.1. Scanner Modification

Modification of the sheet-feed scanner consists of the following steps (the procedure is illustrated in detail in Figure 8):

1. Gain access to the internal micro illumination source (RGB LED) of the scanner.
2. Remove/Disable the internal RGB LED.
3. Structurally modify the scanner housing for placement of a flexible light guide.
4. Connect the external light source to the internal light guide via the flexible light guide.

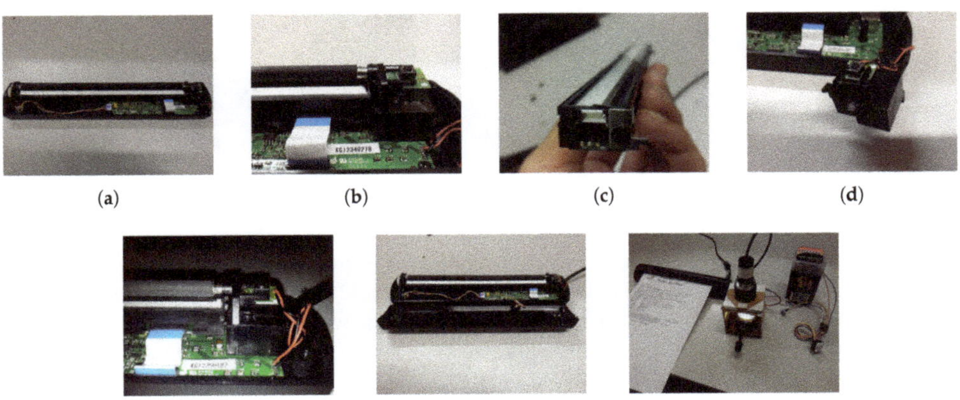

(a) (b) (c) (d)

(e) (f) (g)

Figure 8. The steps of the scanner modification and its connection to the multispectral light source: (**a**) remove the top cover to gain access; (**b**) release the components from the hinge support; (**c**) disengage RGB LED from the scanner sensor; (**d**) make provision for a flexible light guide; (**e**) re-install the components; (**f**) replace the top cover; and (**g**) connect to the multispectral light source.

2.2.2. Scanner Calibration

The multispectral scanner can be calibrated using a special black and white glossy sheet that came with the scanner, a sample of which is shown in Figure 7b. The bright and dark values of each spectral band can be computed using the scanned reference sheet and applying a formula for normalization:

$$C(x,\lambda) = \frac{I(x,\lambda) - D(\lambda)}{B(x,\lambda) - D(x,\lambda)} \tag{1}$$

where I is the original image, C is the calibrated image, and B and D are the average bright and dark points at each wavelength, respectively.

2.3. *Multispectral Document Scanning*

To test the multispectral document scanner, a test page printed from an HP Laserjet Color printer was scanned. The RGB true color image and various bands of a logo in the test page captured by the multispectral scanner are shown in Figure 9. A relative variation in the brightness can be observed between the bands due to the differences in spectral power and bandwidth of the LEDs. The explanation for a relatively darker scanned image using the Amber LED is its low spectral power coupled with a narrow bandwidth, which together cause a weaker response at the detector.

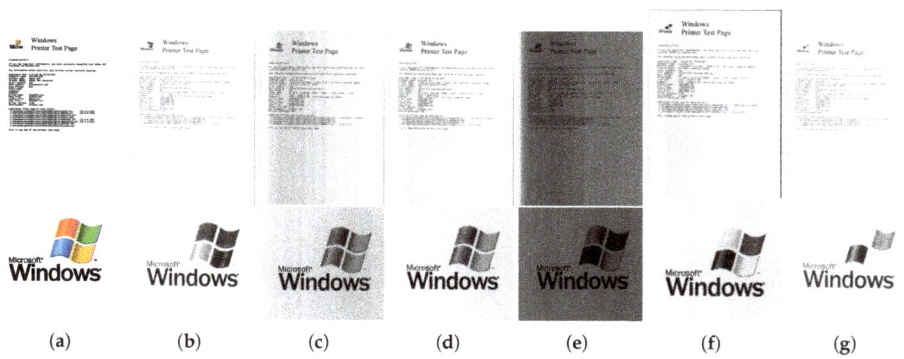

Figure 9. The printed test document (**top**) and a magnified view of the logo contained within (**bottom**) in: (**a**) true RGB; (**b**) royal blue; (**c**) cyan; (**d**) green; (**e**) amber; (**f**) red orange; and (**g**) deep red bands.

Observe that the logo has red, orange, green and blue elements and black text at the bottom. The normalized spectral response computed by averaging an 11 × 11 patch at the center of each colored element is shown in Figure 10. Notice that the different components of the logo have characteristic intensity response to multispectral light according to the spectral band. This demonstrates the ability of the scanner to capture fine details in the spectrum.

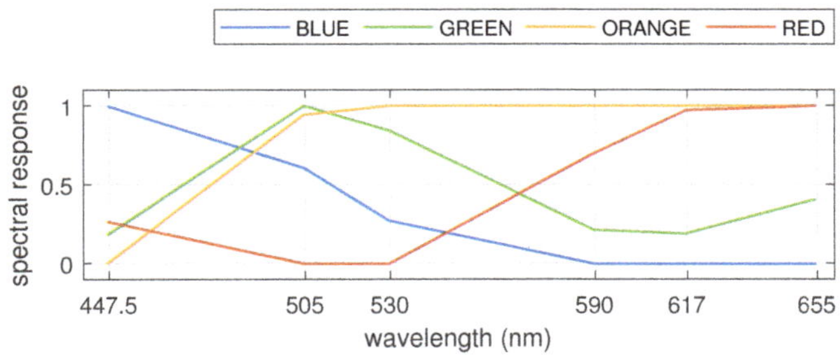

Figure 10. Normalized spectral response of the four colored elements in the Windows logo plotted against the center wavelengths of LEDs in the multispectral light source.

We further tested the developed prototype to identify counterfeit protection system (CPS) codes inserted in color print-outs by all consumer printers [20]. The recent availability of high-resolution printers has not only supported useful purposes, but also paved the way for illegal manipulation of documents. This has consequently persuaded color printer manufacturers to hide an invisible CPS code, which holds information for printer identification. This unique code is printed in every document, in the form of a repeated pattern of yellow dots that is not visible to the naked eye. The unique pattern can be used to identify the source of a document. The multispectral document scanner successfully captures this unique dot-pattern, which can be extracted by binarization of the raw image using image thresholding operation.

The unique patterns of different printers can be identified in terms of their geometrical structures. The two important parameters that form these relationships are the Horizontal Pattern Separation (HPS) distance and the Vertical Pattern Separation (VPS) distance. A raw image of the Royal Blue band of the scanned test page and its patterns enhanced by image processing, which comprised thresholding and image binarization, are shown in Figure 11. The processed image is magnified to visually identify the recurring CPS code. The HPS and VPS measurements were then annotated in the processed image. The CPS code can now be extracted and analyzed using HPS and VPS measurements.

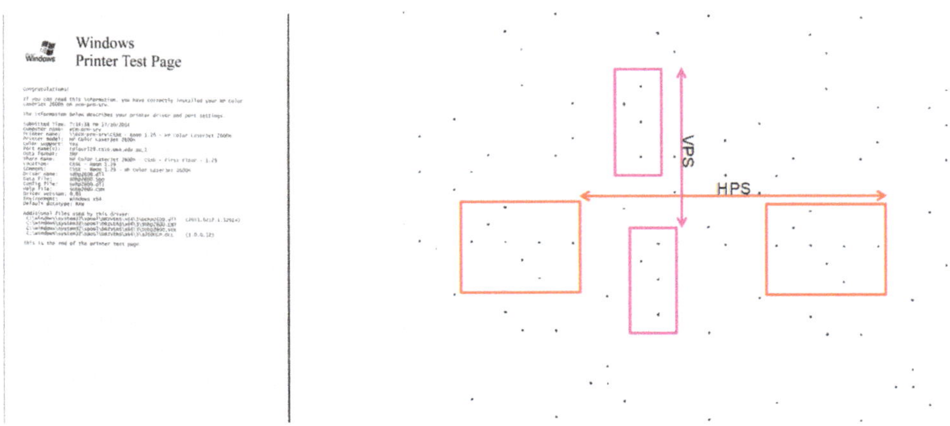

(a) (b)

Figure 11. Extraction of counterfeit protection pattern by image processing: (**a**) Royal Blue band of a test printed document (zoom to 6400% to clearly view the pattern); and (**b**) processed image of enhanced codes separated by horizontal pattern separation (HPS) and vertical pattern separation (VPS).

3. Conclusions

We present the design of a prototype multispectral document scanner, which is demonstrated to capture subtle features in a document using a multispectral light source. The multispectral light source was designed to cover the full range of visible electromagnetic spectrum and connected to a portable sheet-feed document scanner. This light source is comprised of commercial off-the-shelf LEDs of various wavelength, bandwidth and radiant power.

An optimal design may comprise a selection of custom-built LEDs for precise selectivity across the spectrum. These LEDs may also be designed to emit a fixed luminous flux and bring homogeneity in the brightness of bands. The addition of more LEDs will further enhance the capabilities of the device. For instance, the addition of an ultraviolet LED can enable the device to capture invisible security features hidden in some banknotes for verification. Similarly, the addition of an infrared LED can allow the device to capture forgeries in handwritten or printed text for question document examination.

The scanner was calibrated using a white–black reference sheet, which achieved normalization of spectral responses, albeit relative to each band. In circumstances where an absolute spectral response is necessary, the scanner would require calibration with the output of a spectrometer and validated on the same calibration reference. While it is sometimes important to measure absolute spectral response, many applications can simply benefit from a relative (normalized) spectral response measurement to achieve the intended results, as presented in the current system.

In regards to the scanner operation, currently the light source is switched to the desired color by means of a rotary switch, and one band of the document is captured in each pass. A more efficient operation can be built upon electronic switching of the multispectral light source in a time-multiplexed manner to capture all bands in a single feed. However, this modification would require changes to the scanner software to synchronize switching of the external multispectral light-source with the scanning of detector.

Given the presented system and proposed directions of improvement, the prototype design has the potential to be transformed into a fully functional portable device suitable for multipurpose document analysis.

Author Contributions: Z.K. performed the hardware implementations, data curation, experiments and prepared the original written draft; F.S. helped in the experimental design, provided supervision and revised the initial written draft; and A.M. conceptualized the idea, acquired funding, led the complete project and reviewed and edited the final written draft.

Funding: This research work was partially funded by the ARC Grant DP190102443 and the UWA Grant 00609 10300067.

Conflicts of Interest: The authors declare no conflict of interest.

References

1. Leedham, S.S.G. A survey of computer methods in forensic handwritten document examination. In Proceedings of the 11th Conference of the International Graphonomics Society, Scottsdale, AZ, USA, 2–5 November 2003; pp. 278–281.
2. Malik, M.I.; Ahmed, S.; Shafait, F.; Mian, A.S.; Nansen, C.; Dengel, A.; Liwicki, M. Hyper-spectral analysis for automatic signature extraction. In Proceedings of the 17th Conference of the International Graphonomics Society, Pointe-à-Pitre, Guadeloupe, 21–24 June 2015.
3. Baek, S.; Choi, E.; Baek, Y.; Lee, C. Detection of counterfeit banknotes using multispectral images. *Digit. Signal Process.* **2018**, *78*, 294–304. [CrossRef]
4. Lee, J.; Kong, S.G.; Lee, Y.S.; Moon, K.W.; Jeon, O.Y.; Han, J.H.; Lee, B.W.; Seo, J.S. Forged seal detection based on the seal overlay metric. *Forensic Sci. Int.* **2012**, *214*, 200–206. [CrossRef] [PubMed]
5. Saini, K.; Kaur, S. Forensic examination of computer-manipulated documents using image processing techniques. *Egypt. J. Forensic Sci.* **2016**, *6*, 317–322. [CrossRef]

6. Padoan, R.; Steemers, T.; Klein, M.; Aalderink, B.; De Bruin, G. Quantitative hyperspectral imaging of historical documents: Technique and applications. In Proceedings of the International Conference on NDT of Art 2008, Jerusalem, Israel, 25–30 May 2008; pp. 25–30.
7. Khan, M.J.; Yousaf, A.; Khurshid, K.; Abbas, A.; Shafait, F. Automated forgery detection in multispectral document images using fuzzy clustering. In Proceedings of the 13th IAPR International Workshop on Document Analysis Systems (DAS), Vienna, Austria, 24–27 April 2018; IEEE: Piscataway, NJ, USA, 2018; pp. 393–398.
8. Khan, M.J.; Yousaf, A.; Abbas, A.; Khurshid, K. Deep learning for automated forgery detection in hyperspectral document images. *J. Electron. Imaging* **2018**, *27*, 053001. [CrossRef]
9. Shippert, P. Introduction to hyperspectral image analysis. *Online J. Space Commun.* **2003**, *3*, 13.
10. Active Text Ltd. Multi-Spectral Document Scanner. Available online: https://www.active-text.pl/en/multi-spectral-document-scanner (accessed on 24 June 2019).
11. Kim, S.J.; Deng, F.; Brown, M.S. Visual enhancement of old documents with hyperspectral imaging. *Pattern Recognit.* **2011**, *44*, 1461–1469. [CrossRef]
12. Gat, N. Imaging spectroscopy using tunable filters: A review. In *Wavelet Applications VII*; International Society for Optics and Photonics: Bellingham, WA, USA, 2000; Volume 4056, pp. 50–65.
13. Hedjam, R.; Cheriet, M. Historical document image restoration using multispectral imaging system. *Pattern Recognit.* **2013**, *46*, 2297–2312. [CrossRef]
14. Tran, C.D.; Cui, Y.; Smirnov, S. Simultaneous multispectral imaging in the visible and near-infrared region: Applications in document authentication and determination of chemical inhomogeneity of copolymers. *Anal. Chem.* **1998**, *70*, 4701–4708. [CrossRef] [PubMed]
15. Khan, Z.; Shafait, F.; Mian, A. Hyperspectral imaging for ink mismatch detection. In Proceedings of the 12th International Conference on Document Analysis and Recognition, Washington, DC, USA, 25–28 August 2013; IEEE: Piscataway, NJ, USA, 2013; pp. 877–881.
16. Heist, S.; Zhang, C.; Reichwald, K.; Kühmstedt, P.; Notni, G.; Tünnermann, A. 5D hyperspectral imaging: Fast and accurate measurement of surface shape and spectral characteristics using structured light. *Opt. Express* **2018**, *26*, 23366–23379. [CrossRef] [PubMed]
17. Hendargo, H.C.; Zhao, Y.; Allenby, T.; Palmer, G.M. Snap-shot multispectral imaging of vascular dynamics in a mouse window-chamber model. *Opt. Lett.* **2015**, *40*, 3292–3295. [CrossRef] [PubMed]
18. Khan, Z.; Shafait, F.; Mian, A. Automatic ink mismatch detection for forensic document analysis. *Pattern Recognit.* **2015**, *48*, 3615–3626. [CrossRef]
19. Khan, Z.; Shafait, F.; Mian, A. Hyperspectral document imaging: Challenges and perspectives. In Proceedings of the International Workshop on Camera-Based Document Analysis and Recognition, Washington, DC, USA, 23 August 2013; Springer: Berlin/Heidelberg, Germany, 2013; pp. 150–163.
20. Van Beusekom, J.; Shafait, F.; Breuel, T.M. Automatic authentication of color laser print-outs using machine identification codes. *Pattern Anal. Appl.* **2013**, *16*, 663–678. [CrossRef]

sensors

MDPI

Article

An Algorithm Based on Text Position Correction and Encoder-Decoder Network for Text Recognition in the Scene Image of Visual Sensors

Zhiwei Huang [1,2,†], Jinzhao Lin [3,*], Hongzhi Yang [3,†], Huiqian Wang [3], Tong Bai [3], Qinghui Liu [3] and Yu Pang [3,*]

1 School of Communication and Information Engineering, Chongqing University of Posts and Telecommunications, Chongqing 400065, China; hzwnet@swmu.edu.cn

2 School of Medical Information and Engineering, Southwest Medical University, Luzhou 646000, China

3 Chongqing Key Laboratory of Photoelectronic Information Sensing and Transmitting Technology, Chongqing University of Posts and Telecommunications, Chongqing 400065, China; yhz5256@163.com (H.Y.); wanghq@cqupt.edu.cn (H.W.); baitong03@126.com (T.B.); lqh106@163.com (Q.L.)

* Correspondence: linjz@cqupt.edu.cn (J.L.); pangyu@cqupt.edu.cn (Y.P.)

† These authors contributed equally to this work.

Received: 24 April 2020; Accepted: 19 May 2020; Published: 22 May 2020

Abstract: Text recognition in natural scene images has always been a hot topic in the field of document-image related visual sensors. The previous literature mostly solved the problem of horizontal text recognition, but the text in the natural scene is usually inclined and irregular, and there are many unsolved problems. For this reason, we propose a scene text recognition algorithm based on a text position correction (TPC) module and an encoder-decoder network (EDN) module. Firstly, the slanted text is modified into horizontal text through the TPC module, and then the content of horizontal text is accurately identified through the EDN module. Experiments on the standard data set show that the algorithm can recognize many kinds of irregular text and get better results. Ablation studies show that the proposed two network modules can enhance the accuracy of irregular scene text recognition.

Keywords: scene text recognition; visual sensor; text position correction; encoder-decoder network

1. Introduction

The object of natural scene text recognition is to identify the text in the image of natural scene. Natural scene text recognition has important applications in intelligent image retrieval [1,2], license plate recognition [3], automatic driving [4], scene image translation [5] and many other fields.

In recent years, although many effective text recognition methods [6–12] have been proposed and the performance of text recognition has been greatly improved, the text recognition technology of natural scene still has some shortcomings. For the text of natural scene, there is a variety of permutation directions between adjacent texts. In addition to the linear permutation, they may also be arranged in irregular directions such as arcs [13]. For natural scene text arranged in multiple directions, the bounding box may be a rotating rectangle or quadrilateral, so it is difficult to design an effective method to calculate the regularity of the direction of arrangement between adjacent texts [14]. In addition, the irregularity of the visual features of the deformed scene text also hinders the further development of the text recognition technology [15].

The wide variety of text and the diversity of the spatial structure of different types make the visual characteristics of text area have great differences [16], so it is difficult to find a good description feature to classify text area and background area. Therefore, it is also a difficult work to build a multi-classification text recognition framework. Further research is still needed to reach the practical level.

For this reason, we propose a text recognition algorithm based on TPC-EDN to realize a better recognition of various types of irregular text in natural scenes. The algorithm uses TPC module to modify the slanted text into horizontal text for easy recognition, and then accurately identifies the text content through EDN model. The encoder network (EN) module uses dense connection network and BLSTM to effectively extract the spatial and sequence characteristics of text and generate coding vectors. The decoder network (DN) module converts the encoding vector into the output sequence through the attention mechanism and LSTM.

Our contributions in this paper are as follows: First, we propose a TPC approach which is a coordinate offset and regression method based on CNN to realize the end-to-end training. Second, we introduce EN module to extract text features based on dense connection network and BLSTM. Third, the training process of our proposed algorithm is simple and fast, and it is robust to irregular text recognition.

2. Overall Network Structure

The text recognition algorithm designed in this paper mainly includes two modules: the text position correction module and the encoder-decoder network module. The TPC module corrects the detected oblique text into horizontal text, then the EDN module recognizes horizontal text. EDN module includes the encoder network (EN) and the decoder network (DN). The EN uses the dense block and two-layer BLSTM [17] methods to extract text features, and can generate feature vector sequences with character context feature relations. The DN uses the attention mechanism [18] to weight the encoded feature vectors, which can make more accurate use of character-related information. Then, through a layer of LSTM [19], DN adopts the output of the previous moment and the input of the current moment to jointly determine the recognition result of the current moment. The overall structure is shown in Figure 1.

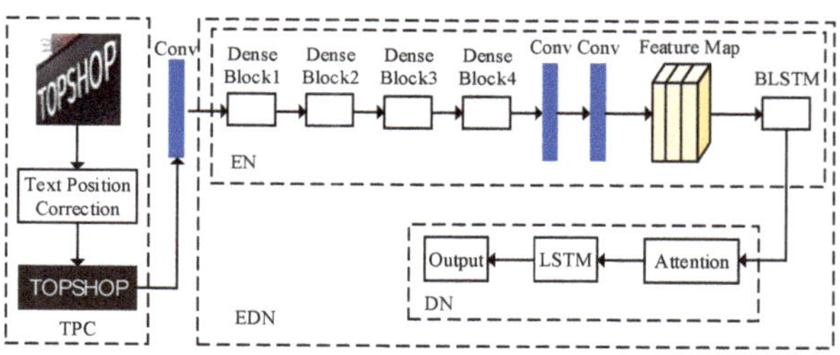

Figure 1. Overall structure of our text recognition algorithm.

2.1. Text Position Correction Module

TPC is the main research method for the oblique text recognition, which corrects the oblique text into the horizontal text, and then carries on the recognition to the horizontal text. Most of the traditional text position correction methods are based on affine transformation [20], which has good effects on text with small tilt angle, but bad effects on text with large tilt angle and are difficult to train. In the study of text recognition algorithm, this paper proposes an improved TPC method based on the idea of variable convolution two-dimensional offset [21] and offset sampling [22], which is a coordinate offset regression method based on CNN. It can be combined with other neural networks to complete end-to-end training, and the training process is simple and fast. The detailed structure is shown in Figure 2.

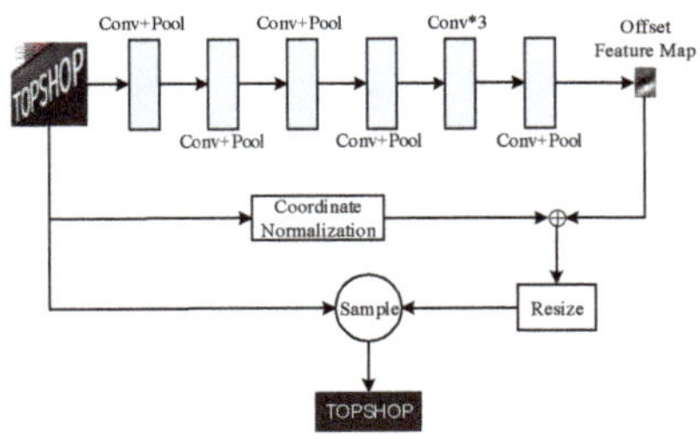

Figure 2. TPC structure diagram.

As can be seen from Figure 2, the TPC process of this paper is as follows: Firstly, a pre-processing step is carried out to process the input text into the same size, which can speed up the training process of the algorithm. Secondly, the spatial features of pixels [23] are extracted by CNN to obtain a fixed size feature map, in which each pixel corresponds to a part of the original image. This is equivalent to splitting the original image into several small pieces, and the prediction of coordinate offset for each piece is the same as the two-dimensional offset prediction of the deformable convolution. Thirdly, the offset is then superimposed on the normalized coordinates of the original image. Finally, the Resize module uses a bilinear interpolation method to sample the text feature map to the original size as the revised text.

The input of the whole text recognition algorithm is the text bounding box detected by the text detection algorithm. Due to the irregular shape of the text, the size of the detected text bounding box is different. If the text is directly input into the text recognition algorithm, the training speed of the text recognition algorithm will be reduced. Therefore, after the preprocessing module, the text bounding box is fixed to a uniform size, namely 64 pixels in height and 200 pixels in width, and then the feature map is obtained by extracting features continuously through CNN, and the coordinate offset is returned. The detailed structure and parameter configuration of TPC are shown in Table 1.

Table 1. Detailed structure of TPC module.

Type	Configuration	Size
The Input	-	$1 \times 64 \times 200$
Conv	k3, num64, s1, p1	$64 \times 64 \times 200$
AVGPool	k2, s2	$64 \times 32 \times 100$
Conv	k3, num128, s1, p1	$128 \times 32 \times 100$
AVGPool	k2, s2	$128 \times 16 \times 50$
Conv	k3, num256, s1, p1	$256 \times 16 \times 50$
AVGPool	k2, s2	$256 \times 8 \times 25$
Conv	k3, num128, s1, p1	$128 \times 8 \times 25$
AVGPool	k2, s1	$128 \times 7 \times 24$
Conv	k3, num64, s1, p1	$64 \times 3 \times 12$
Conv	k3, num32, s1, p1	$32 \times 3 \times 12$
Conv	k3, num8, s1, p1	$8 \times 3 \times 12$
Conv	k3, num2, s1, p1	$2 \times 3 \times 12$
AVGPool	k2, s1	$2 \times 2 \times 11$
Tanh	-	$2 \times 2 \times 11$
The Resize	-	$2 \times 64 \times 200$

In Table 1, k3 means the size of convolution kernel is 3×3, num64 means the number of convolution kernel is 64, s1 means the stride is 1, p1 means the padding is 1, Conv means convolution, and AVGPool means average pooling. The number of convolution kernels gradually increases from the first layer, and then decreases. Finally, the number is set as 2 in order to generate a two-dimensional offset feature map, whose size is 2×11. This is equivalent to dividing the entire input image into 22 blocks, each corresponding to the corresponding coordinate offset value. The activation function Tanh is used to adjust the predicted value of the migration to between [−1, 1], and return the offset of the X-axis and the offset of the Y-axis through two channels respectively. Then, the Resize module is used to sample the offset feature map of the two channels to the size of the original figure 2 × 64 × 200. Sample is a bilinear interpolation up-sampling module to obtain the revised text.

Each value in the offset feature map represents its corresponding coordinate offset of the point in the original image. In order to correspond to the dimension of the feature map, the coordinates of each pixel in the original image need to be normalized. The normalized coordinate interval is between [−1, 1], and it also contains two channels, namely the X-axis channel and Y-axis channel. Figure 3 is the comparison of the original image before and after the normalization of coordinates.

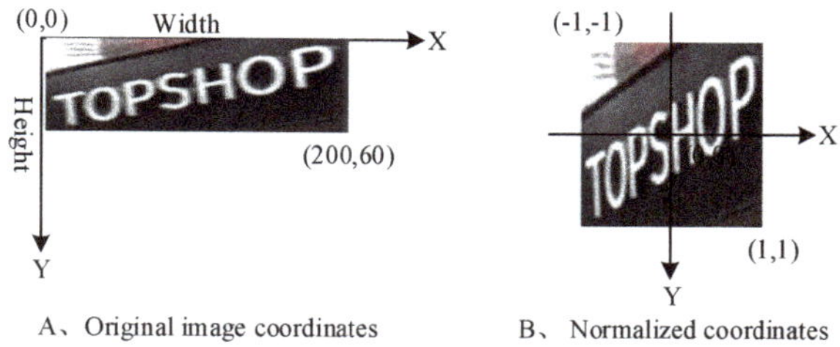

Figure 3. Schematic diagram of coordinate normalization.

The image is stored in the form of matrix in the computer, so the upper left corner of the image in Figure 3 is the origin of the coordinate axis (0,0), the horizontal axis represents the width of the image, and the vertical axis represents the height of the image. After normalization, the center of the image is the origin of the coordinate, the upper left corner in Figure 3 is the coordinate (−1,−1), and the lower right corner is the coordinate (1,1). The generated normalized image is double-channel, and the coordinates of pixels in the same position on different channels are the same. After that, the offset feature image is superimposed with the corresponding area of the normalized image to complete the correction of the corresponding position of each pixel. The formula is expressed as:

$$T_{(channel,i,j)} = offset_{(channel,i,j)} + G_{(channel,i,j)}, channel = 1,2 \quad (1)$$

$$F'_{(ii',jj')} = F(i',j') \quad (2)$$

where, *channel* refers to the number of channels, T represents the feature map after position correction, *offset* represents the offset feature map, G represents the normalized image, (i, j) represents the coordinates of the normalized image, (i', j') represents the coordinates of the original image, (ii', jj') represents the revised offset coordinates, F' represents the corrected image, F represents the original image.

Adding the corresponding offset to the normalized image, the offset of each point coordinate on the normalized image occurs in both horizontal and vertical directions. The offset is (Δx, Δy), the revised offset coordinate is (ii, jj), and then the size is up-sampled to the original size by bilinear interpolation method. The revised image F' is obtained, whose corresponding coordinate is (ii', jj'), The relation between the original image and the normalized image is shown in Formula (2). The pixels of the two points remain the same size, just the position coordinates are changed.

2.2. Encoder Network

The EN module encodes the spatial and sequential features of extracted text images into fixed feature vectors [24]. Feature extraction network [25] plays a key role in the EN module. A good feature extraction network can determine the quality of encoding and has a great impact on the recognition effect of the whole text recognition algorithm. In this paper, the EN module adopts the methods of dense connection network and BLSTM to extract text features, in which dense connection network can extract rich spatial features of text images. Considering the context sequence feature of text, the feature relation between different characters can be learned by BLSTM. The EN module designed in this paper is easy to train and has a good effect. A brief introduction about it is as follows:

(1) Dense connection network is stacked by several dense blocks. Taking the advantages of DenseNet [26] in feature extraction, dense connection network is used to improve the direction of information flow during feature extraction, and all layers in a dense block can be connected by jumping. Each convolution layer can obtain feature information from all previous layers, enhance the reuse of multi-layer features, and transmit feature information to all subsequent layers. At the same time, the method of jumping direct connection makes it easier to obtain the gradient in the process of back propagation, simplifies the feature learning process, and alleviates the gradient dispersion problem.

(2) The detailed structure of the two BLSTMs is shown in Figure 4. Each BLSTM has two hidden layers, recording two states of the current time t: one is the forward state from front to back, the other is the reverse state from back to front. The input of the first layer is the sequence of feature vectors extracted by CNN $\{x_0, x_1, \ldots, x_i\}$, the output after a layer of BLSTM is $\{y_0^1, y_1^1, \ldots, y_i^1\}$. And then it is taken as the input of the second layer, finally the output sequence $\{y_0^2, y_1^2, \ldots, y_i^2\}$ can be got. As can be seen from Figure 4, the output of each time t is determined by the hidden layer state in both directions. In this paper, two BLSTMs are stacked to learn the feature states of the four hidden layers, which can not only store more memory information, but also better learn the relationship between feature vectors.

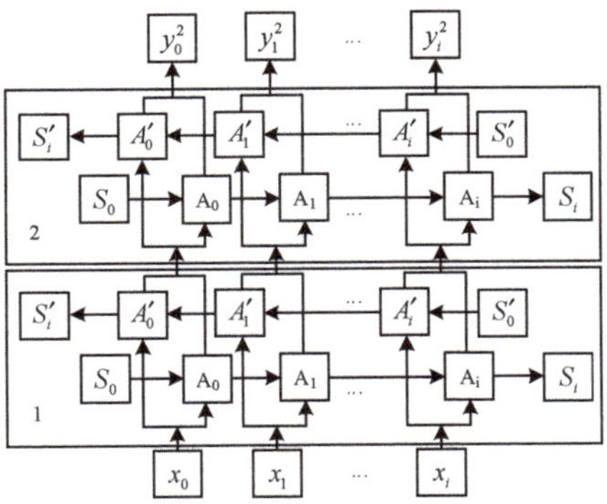

Figure 4. BLSTM structure diagram of two floors.

The dense block generates a two-dimensional feature map, while the input of BLSTM is in the serialized form. Therefore, it is necessary to convert the feature map into the sequence form of feature vectors, and then learn the context feature relationship between sequences through BLSTM. Figure 5 shows the process of transforming the feature map into the feature vector sequence. The feature map is evaluated according to the column of a certain width, the vertical direction is taken as a feature vector.

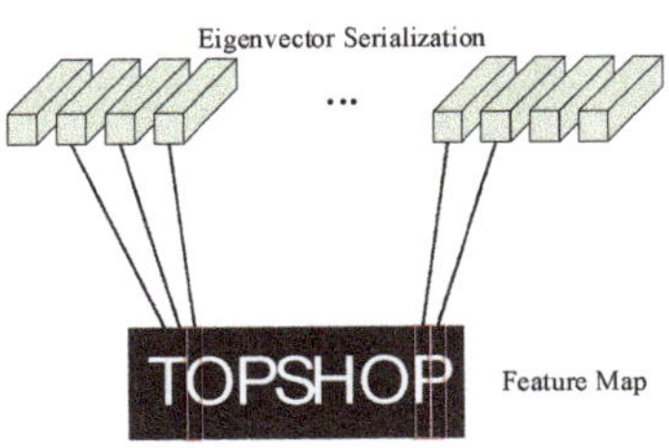

Figure 5. Feature map is transformed into feature vector sequence.

As can be seen from Figure 5, the character "O" requires multiple feature vectors to determine the output value, and it is impossible to accurately predict the character by relying on only one feature vector. Therefore, learning the correlation between feature vectors through BLSTM plays an important role in character recognition.

The EN module adopts four dense blocks, followed by two convolution layers, between which there is one Max Pooling and activation function layer, and then two BLSTM layers. The detailed parameters of the EN module are shown in Table 2.

Table 2. Detailed parameters of EN module.

Type	Configuration
Convolution	[k3, num32, s1, p1]
Max Pooling	[k2, s1]
The Activation Function	Swish
Dense Block	[k3, num32, s1, p1] x 4
Dense Block	[k3, num64, s1, p1] x 4
Dense Block	[k3, num128, s1, p1] x 4
Dense Block	[k3, num256, s1, p1] x 4
Convolution	[k3, num128, s1, p1]
Max Pooling	[k2, s1]
The Activation Function	Swish
Convolution	[k3, num128, s1, p1]
BLSTM	Hidden unit: 256
BLSTM	Hidden unit: 256

As can be seen from Table 2, EN module adopts several convolution layers, pooling layers and activation function layers. The detailed parameters of the convolution layer include the size of convolution kernel, the number of convolution nuclei, stride and padding, which are respectively represented by k, num, s and p. The Max Pooling method is adopted in the all pooling layers, and the parameters are convolution kernel size k and stride s. The activation function takes the Swish function. The number of convolution nuclei in the four dense blocks gradually increases. In each Dense Block, "×4" represents four consecutive convolution layers, followed by two convolution operations. Finally, two BLSTMs are adopted, in which the number of hidden layer units of BLSTM in each layer is 256.

2.3. Decoder Network

The DN module is the reverse process of the EN module, which decodes the encoded feature vectors into output sequences and makes the decoding state as close as possible to the original input state. The text area of a text image usually exists in the form of a sequence, with variable length, and its feature vector is serialized. Therefore, this paper adopts soft attention mechanism [27] to focus the serialized feature vectors according to the weight distribution, which can effectively use the character features at different moments to predict the output value, and finally connects a layer of LSTM, which can store the past state and determine the output of the current moment through the output of the previous moment. The detail structure on DN is shown in Figure 6.

Figure 6 shows that the feature vector sequence generated by the EN is directly used as the input of the DN, the hidden layer of BLSTM in the process of the EN contains context feature of text feature vector sequence, the feature vector set can be set as $[h_1, h_2, \ldots, h_i, \ldots, h_T]$, in which the feature Hi generated at each moment i consists of two directions of feature combination, $h_i = [h_i, h_i^*]$. C_t is the semantic encoding vector of the attention model, represents the weighted value of hidden layer feature hi at time t in LSTM network, is expressed as Equation (3).

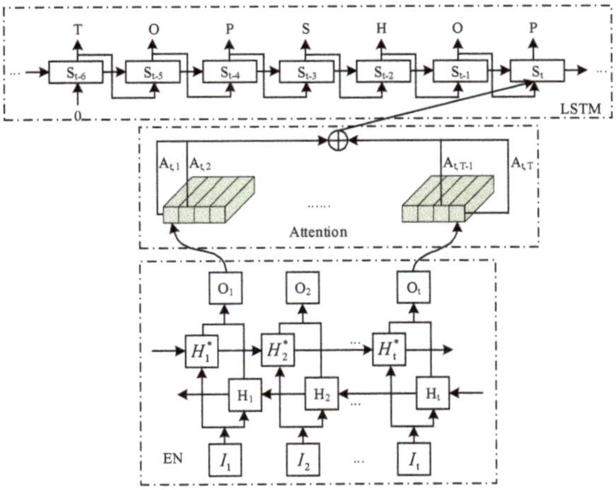

Figure 6. Detail structure on Decoder Network.

In Figure 6, T represents the attention range of the attention mechanism, and its length is 30. If T is too large, the hidden layer needs to remember too much information, the calculation of the model increases rapidly, and the general text statement rarely exceeds 30 words. And too large T value will also make the model's attention be distracted, so that the DN module cannot focus on the key feature vectors, and the decoding effect is not good. In this paper, the designed DN module takes the predicted output of the previous moment as the input of the current moment through LSTM, which can serve as a reference for the prediction of the current moment. In Figure 6, the output at the current moment can be accurately determined to be "P" based on the past output state. The detailed Formula (3)–(7) of the whole decoding process is as follows:

$$C_t = \sum_{i=1}^{T} A_{t,i} h_i \tag{3}$$

$$A_{t,i} = \frac{\exp(e_{t,i})}{\sum_{k=1}^{T} \exp(e_{t,k})} \tag{4}$$

$$e_{t,i} = f_{att}(s_{t-1}, h_i) \tag{5}$$

$$s_t = f(s_{t-1}, y_{t-1}, C_t) \tag{6}$$

$$y_t = g(y_{t-1}, s_t, C_t) \tag{7}$$

In the above Equations (3)–(7), $A_{t,i}$ represents the attention weight after normalization, $e_{t,i}$ represents the weight of attention, s_{t-1} represents the hidden layer state of the DN module at time $t-1$, s_t represents the hidden layer state of the DN module at time t, f and g represent the nonlinear activation function, and y_t represents the predicted output of the DN module at time t. y_t is determined by the predicted output y_{t-1} of the previous moment, the hidden layer state s_t of the DN module and the attention semantic coding C_t.

3. Implementation Details

All experiments with the text recognition algorithm in this paper are completed in the PyTorch framework. The experimental workstation is equipped with a 3.6 GHz Intel i7-6800k CPU, 64G RAM, eight GTX 2080Ti GPUs, and the operating system is Ubuntu 16.04. In the training process, CUDA 9.0 and Cudnn 7.1 are adopted for GPU acceleration, which can significantly improve the training speed. OpenCV 3.2 with Python 3.6 is used to visualize the results. The parameter settings used in the training process are shown in Table 3.

Table 3. Parameter settings

Type	Configuration
The input size	64 × 200
The iterations	100,000
Batch size	16
Learning rate	10^{-3}
Learning rate attenuation	0.9/10,000 of the iterations

4. Experiments

4.1. Experimental Data Set

The data set used in the text recognition algorithm is different from the text detection, which is usually more standard, multilingual and simple. In order to verify the advantages of the text recognition algorithm in this paper, we conducted experimental comparison on a variety of data sets, using the data sets such as SVT, ICDAR 2013, IIIT5K-Words and CUTE80. The following is a detailed introduction. The sample of the scene texts in the data sets in this paper is seen in Figure 7.

SVT ICDAR 2013 IIIT5K CUTE80

Figure 7. Sample of the scene texts in the data sets in this paper.

SVT [28]: this data set comes from Google Street View. The text size is diverse, the text direction is not fixed, many pictures are polluted by noise and mixed background, and the image resolution is low. This data set can effectively test the text recognition ability of the text recognition algorithm. It contains 647 cropped images and used the two common data formats: SVT-50, SVT-None. "50" means that the annotated dictionary library contains 50 words, and "None" means that there is no dictionary library. The same is true for the following data set.

ICDAR 2013 [29]: this data set is a commonly used data set for text recognition. The text in the image is usually horizontal and the text background is simple. The image format of this data set is the same as that of ICDAR 2003 [30], including 1015 cropped images. The following three data formats are commonly used: ICDAR 2013-50, ICDAR 2013-FULL and ICDAR 2013-None. Each image in this data set has a complete ground truth.

IIIT5K [31]: this data set is collected on the Internet and contains 3000 cropped images. It is a commonly used horizontal text data set. There are three commonly used data formats: III5K-50 with 50 annotated words, III5K-1k with 1000 annotated words, and III5K-None with no annotated words. Each image in this data set has a complete ground truth.

CUTE80 [32]: this data set is a commonly used slanted text or curved text data set, mainly used to evaluate the recognition effect of the algorithm model on multi-direction slanted text and curved text. It contains 288 clipped images and is a data set without dictionary annotation.

4.2. Experimental Results and Analysis

In order to verify the effect of the text recognition algorithm and the influence of each sub-module on the text recognition results, the experimental analysis was carried out on each sub-module, and the experimental verification was carried out on the whole text recognition scheme. And in order to validate the importance of each sub-module, we carried on the ablation study [33] in this article. Firstly, we removed the sub-module and test the whole text recognition algorithm, and then added the module and conducted comparison experiment on the whole text recognition algorithm. If there is no significant improvement in the accuracy of text recognition after adding the module, the module will be removed to simplify the algorithm architecture.

As many references use recognition accuracy to evaluate the text recognition algorithm, in order to compare with other text recognition algorithms, this paper adopts recognition accuracy and training time as the evaluation criteria. The following is the detailed experimental results and analysis of the text recognition algorithm.

4.2.1. TPC and its Influence on Text Recognition Results

The TPC module corrects the tilted text into horizontal text through coordinate offset, and uses the EDN module to recognize the horizontal text. In this paper, the importance of the TPC module can be verified by ablation study. The experimental data sets are SVT and IIIT5K, the setting of training parameters is shown in Table 3, the comparison of experimental results is shown in Table 4. In order to demonstrate the effect of TPC module, this experiment selects three images to test the text recognition algorithm. These images all have the characteristics of blur, tilt and bending so as to verify the effect of the text recognition algorithm modified by TPC module.

Table 4. TPC's recognition accuracies (%) and its ablation study.

Model	SVT			IIIT5K			
	50	None	Time	50	1 k	None	Time
Without TPC	89.6	76.5	3.2 h.	93.2	92.5	81.6	6.1 h.
With TPC	96.5	83.7	3.6 h.	99.4	98.1	88.3	6.7 h.

From the experimental result in Table 4, it can be seen that in the data set SVT the recognition accuracy is significantly improved by more than 6%, which indicates that the text recognition algorithm can more accurately identify the slanted text content after using TPC module to correct the text position. The recognition accuracy in data set IIIT5K is also greatly improved, which indicates that it is also suitable for normal horizontal text. TPC module will increase the training time of the whole model during the training process, but the increase is relatively small and has little impact on the performance.

4.2.2. Dense Connection Network and Its Impact on Text Recognition Results

Dense connection network is an important part of EN module. In order to verify the influence of the network on the whole text recognition algorithm, ablation experiments were carried out. The experimental data sets were ICDAR2013 and IIIT5K. Experimental results are shown in Table 5.

Table 5. Recognition accuracies (%) of dense connection network and its ablation study.

Model	ICDAR 2013				IIIT5K			
	50	FULL	None	Time	50	1 k	None	Time
Without DCN	89.6	86.5	79.5	5.4 h.	91.2	89.5	80.5	6.5 h.
With DCN	98.6	97.5	92.3	5.5 h.	99.4	98.1	88.3	6.7 h.

It can be seen from Table 5 that the dense connection network has a great impact on the whole text recognition algorithm and can significantly improve the accuracy of text recognition. In the data sets ICDAR2013 and IIIT5K, with or without dictionary annotation, the accuracy of text recognition is improved by more than 7%, and even reaches 99.4% in IIIT5K-50, which indicates that dense connection network can effectively improve the recognition effect of the text recognition algorithm. After adding the dense connection network, the training time of the model increases little, only about 0.2 h, the result indicates that the dense connection network can improve the back propagation process of the neural network and has a certain optimization effect on the training process of the whole model.

4.2.3. Depth of BLSTM in EN and Its Influence on the Text Recognition Results

This experiment verifies the influence of different depth of BLSTM on the text recognition results. The depth of BLSTM may affect the feature learning ability of the text recognition algorithm. A certain depth can be used to learn more sequence features, but the continuous increase of depth will increase the amount of parameter calculation, result in a longer training time. Therefore, through the experimental comparison of BLSTM with different depth, it is necessary to select the appropriate depth from the accuracy and training time. The experimental data sets are ICDAR2013 and IIIT5K. For the convenience of comparison, the training time is the average time of the text recognition algorithm in three different structures of each data set, and the unit is hour. The setting of training parameters is shown in Table 3, the experimental results are shown in Table 6.

Table 6. Depth of BLSTM in EN and its recognition accuracies (%).

Depth	ICDAR 2013				IIIT5K			
	50	FULL	None	Time	50	1 k	None	Time
0	89.6	89.5	83.5	3.4 h.	91.2	90.5	82.6	4.5 h.
1 layer	97.2	94.6	89.9	4.7 h.	93.6	93.3	85.1	5.6 h.
2 layers	98.6	97.5	92.3	5.5 h.	99.4	98.1	88.3	6.7 h.
3 layers	98.3	97.1	92.2	6.8 h.	99.3	97.8	88.1	7.9 h.

It can be seen from Table 6 that in ICDAR2013-50 the recognition accuracy of BLSTM in the first layer was improved by 7.6% when compared with BLSTM with without BLSTM, which indicates that BLSTM can improve the recognition accuracy of the text recognition algorithm. In addition, as the number of BLSTM layers increases, the recognition accuracy of the algorithm gradually increases and the training time also increases. When the number of BLSTM layers is 2, the recognition accuracy of the text recognition algorithm reaches the highest, which reaches 98.6%, 97.5% and 92.3% in three data formats of ICDAR2013, and reaches 99.4%, 98.1% and 88.3% in three data formats of IIIT5K, respectively. When the number of layers of BLSTM is increased, the recognition accuracy does not increase, but decreases. It is analyzed that the algorithm is too complex, which leads to the over-fitting phenomenon in the nonlinear learning process. At the same time, the number of parameters of the algorithm increases and the training time becomes longer, which is a great challenge to the hardware equipment. Therefore, the two-layer BLSTM is the most reasonable choice in this paper.

4.2.4. Attention Mechanism in DN and Its Influence on Text Recognition Results

Adding attention mechanism to DN can make use of feature information reasonably and improve the decoding efficiency of feature effectively. The effect of attention mechanism on text recognition algorithm was verified by ablation study. The experimental data sets are SVT and IIIT5K, and the setting of training parameters is shown in Table 3, the experimental results are shown in Table 7.

Table 7. Attention mechanism and its recognition accuracies (%).

Model	SVT			IIIT5K			
	50	None	Time	50	1 k	None	Time
Without Attention	92.7	80.5	3.4 h.	95.5	93.5	84.6	6.5 h.
With Attention	96.5	83.7	3.6 h.	99.4	98.1	88.3	6.7 h.

It can be seen from Table 7 that the recognition accuracy with attention mechanism in SVT and IIIT5K can both improved by more than 3%. It shows that the attention mechanism can extract the character effectively in the text recognition algorithm, which is very important to improve the effect of text recognition. At the same time, the training time of the whole model increases by only 0.2 h after the attention mechanism is added, which indicates that the model complexity of the attention mechanism is low and the parameter calculation amount of the whole model does not increase significantly.

4.3. *Results Compared with Other Text Recognition Algorithms*

The above experiments are conducted on the current algorithms for text recognition, and prove the validity of our algorithm in this paper. To validate the effect on tilted text recognition of our text recognition algorithm, we select SVT and CUTE80 as the experimental data sets. The training parameters are shown in Table 3, the experimental results are shown in Table 8.

Table 8. Recognition accuracies (%) compared with other text recognition algorithms.

Model	SVT-50	SVT-None	CUTE80
Bissacco et al. [34]	90.4	78.0	-
He et al. [35]	95.4	80.7	-
Jaderberg et al. [36]	93.2	71.7	42.7
Lee et al. [37]	96.3	80.7	-
Shi et al. [38]	96.4	80.8	54.9
Shi et al. [39]	95.5	81.9	59.2
Yang et al. [40]	95.2	-	69.3
Ours	96.5	83.7	71.3

From the experimental data in Table 8, it can be seen that the text recognition algorithm in this paper can achieve a good recognition accuracy even in the data sets annotated by different dictionaries, no matter the text is tilted or curved. In the two data formats SVT-50 and SVT-None, the text recognition accuracy reached 96.5% and 83.7% respectively. In the curved text data set CUTE80, the recognition accuracy reached 71.3%, indicating that the text recognition algorithm designed in this paper has good robustness for the recognition of slanted and curved text. Because some algorithms do not carry out text recognition experiments in the specified data set, there is represented by "-" in Table 8.

4.4. *Experimental Results of Text Recognition in the Scene Image of Visual Sensors*

For the overall scheme of text recognition in natural scenes captured by visual sensors, the text recognition algorithm is combined with the text detection algorithm for experiments, and a demonstration is designed to directly identify the text images in natural scenes. Figure 8 shows the results.

The test images in Figure 8 were all randomly collected in the natural scene by visual sensors. The image background is complex, the text size is variable, and the text direction is skewed. The red box in the left column is the detection result of the text detection algorithm, while the right column is the recognition result of the text recognition algorithm. According to the recognition results in the right column of Figure 8, the natural scene text recognition algorithm designed in this paper can accurately identify the text in the figure, and has a good recognition effect for both multi-scale text and inclined text, indicating that the algorithm is feasible.

Figure 8. Experimental results of text recognition in a visual sensor scene image.

5. Conclusions

In this paper, a text recognition algorithm based on TPC-EDN is proposed to solve the problem that text recognition algorithm only with encoding and decoding network is not good for oblique text recognition. Firstly, we analyze the problems existing in the current text position correction method, put forward the TPC method in this paper. Secondly, we consider the text sequence feature, design a text recognition algorithm based on EDN, and adopt the attention mechanism to effectively improve decoding accuracy. Finally, we test the text recognition algorithm and conduct ablation study, compare some experimental data and verify the advantage of our algorithm.

Author Contributions: Z.H. analyzed the results and wrote the manuscript. H.Y. designed and the prototype, implemented the experiments. H.W. and T.B. funded the research. J.L., Q.L. and Y.P. reviewed the manuscript. All authors have read and agreed to the published version of the manuscript.

Funding: This work was funded by National Natural Science Foundation of China (61671091, 61971079), by Science and Technology Research Program of Chongqing Municipal Education Commission (KJQN201800614), by Chongqing Research Program of Basic Research and Frontier Technology (cstc2017jcyjBX0057, cstc2017jcyjAX0328), by Scientific Research Foundation of CQUPT(A2016-73), by Key Research Project of Sichuan Provincial Department of Education (18ZA0514) and by Joint Project of LZSTB-SWMU(2015LZCYD-S08(1/5)).

Conflicts of Interest: The authors declare no conflict of interest.

References

1. Zhu, L.; Shen, J.; Xie, L. Unsupervised topic hypergraph hashing for efficient mobile image retrieval. *IEEE Trans. Cybern.* **2017**, *47*, 3941–3954. [CrossRef] [PubMed]
2. Piras, L.; Giacinto, G. Information fusion in content based image retrieval: A comprehensive overview. *Inf. Fusion* **2017**, *37*, 50–60. [CrossRef]
3. Bulan, O.; Kozitsky, V.; Ramesh, P. Segmentation-and annotation-free license plate recognition with deep localization and failure identification. *IEEE Trans. Intell. Transp. Syst.* **2017**, *18*, 2351–2363. [CrossRef]
4. Islam, K.T.; Raj, R.G.; Mujtaba, G. Recognition of traffic sign based on bag-of-words and artificial neural network. *Symmetry* **2017**, *9*, 138. [CrossRef]
5. Wang, X.J.; Zhang, L.; Ma, W.Y. Text to Image Translation. US Patent 9678992B2, 13 June 2017.
6. Jaderberg, M.; Simonyan, K.; Zisserman, A.; Kavukcuoglu, K. Spatial Transformer Networks. Available online: https://arxiv.org/abs/1506.02025 (accessed on 20 May 2020).
7. Zhan, F.; Lu, S. ESIR: End-to-end Scene Text Recognition via Iterative Image Rectification. In Proceedings of the the IEEE Conference on Computer Vision and Pattern Recognition (CVPR), Los Angeles, CA, USA, 16–19 June 2019; pp. 2059–2068.

8. Shi, B.G.; Yang, M.K.; Wang, X.G. ASTER: An Attentional Scene Text Recognizer with Flexible Rectification. *IEEE Trans. Pattern Anal. Mach. Intell.* **2018**, *41*, 2035–2048. [CrossRef]
9. Su, B.L.; Lu, S.J. Accurate Scene Text Recognition Based on Recurrent Neural Network. In Proceedings of the Asian Conference on Computer Vision, Singapore, 1–5 November 2014; pp. 35–48.
10. Su, B.L.; Lu, S.J. Accurate recognition of words in scenes without character segmentation using recurrent neural network. *Pattern Recognit.* **2017**, *63*, 397–405. [CrossRef]
11. Yu, C.; Song, Y.; Zhang, Y. Scene text localization using edge analysis and feature pool. *Neurocomputing* **2016**, *175*, 652–661. [CrossRef]
12. Chng, C.K.; Chan, C.S. Total-Text: A Comprehensive Dataset for Scene Text Detection and Recognition. In Proceedings of the 2017 14th IAPR International Conference on Document Analysis and Recognition (ICDAR), Kyoto, Japan, 9–15 November 2017. [CrossRef]
13. Ma, J.; Shao, W.; Ye, H.; Wang, L.; Wang, H.; Zheng, Y.; Xue, X. Arbitrary-Oriented Scene Text Detection via Rotation Proposals. *IEEE Trans. Multimed.* **2018**, *20*, 3111–3122. [CrossRef]
14. Ren, X.; Zhou, Y.; Huang, Z.; Sun, J.; Yang, X.; Chen, K. A Novel Text Structure Feature Extractor for Chinese Scene Text Detection and Recognition. *IEEE Access* **2017**, *5*, 3193–3204. [CrossRef]
15. Tang, Y.; Wu, X. Scene Text Detection Using Superpixel-Based Stroke Feature Transform and Deep Learning Based Region Classification. *IEEE Trans. Multimed.* **2018**, *20*, 2276–2288. [CrossRef]
16. Tian, S.; Yin, X.C.; Su, Y.; Hao, H.W. A Unified Framework for Tracking Based Text Detection and Recognition from Web Videos. *IEEE Trans. Pattern Anal. Mach. Intell.* **2018**, *40*, 542–554. [CrossRef] [PubMed]
17. Huang, Z.; Xu, W.; Yu, K. Bidirectional lstm-crf Models for Sequence Tagging. Available online: https://arxiv.org/abs/1508.01991 (accessed on 20 May 2020).
18. Gao, Y.T.; Huang, Z.; Dai, Y.C.; Xu, C.; Chen, K.; Guo, J. Double Supervised Network with Attention Mechanism for Scene Text Recognition. Available online: https://arxiv.org/pdf/1808.00677 (accessed on 20 May 2020).
19. Hochreiter, S.; Schmidhuber, J. Long Short-Term Memory. *Neural Comput.* **1997**, *9*, 1735–1780. [CrossRef] [PubMed]
20. Jaderberg, M.; Simonyan, K.; Zisserman, A. Spatial transformer networks. *Neural Inf. Process. Syst.* **2015**, *20*, 2017–2025.
21. Dai, J.F.; Qi, H.Z.; Xiong, Y.W. Deformable convolutional networks. In Proceedings of the International Conference on Computer Vision, Venice, Italy, 22–29 October 2017; pp. 764–773.
22. Kobchaisawat, T.; Chalidabhongse, T.H.; Satoh, S.I. Scene Text Detection with Polygon Offsetting and Border Augmentation. *Electronics* **2018**, *9*, 117. [CrossRef]
23. Tao, P.; Yi, H.; Wei, C. A method based on weighted F-score and SVM for feature selection. In Proceedings of the 2013 25th Chinese Control and Decision Conference (CCDC), Guiyang, China, 25–27 May 2013; pp. 4287–4290.
24. Sutskever, I.; Vinyals, O.; Le, Q.V. Sequence to Sequence Learning with Neural Networks. *Neural Inf. Process. Syst.* **2014**, *8*, 3104–3112.
25. Lin, T.; Dollar, P.; Girshick, R.B.; He, K.; Hariharan, B.; Belongie, S. Feature Pyramid Networks for Object Detection. *Comput. Vis. Pattern Recognit.* **2017**, *6*, 936–944.
26. Huang, G.; Liu, Z.; Van Der Maaten, L. Densely Connected Convolutional Networks. *Computer Vis. Pattern Recognit.* **2017**, *12*, 2261–2269.
27. Vaswani, A.; Shazeer, N.; Parmar, N. Attention is all you need. *Neural Inf. Process. Syst.* **2017**, *14*, 5998–6008.
28. Wang, K.; Babenko, B.; Belongie, S. End-to-end scene text recognition. In Proceedings of the 2011 International Conference on Computer Vision, Barcelona, Spain, 6–13 November 2011; pp. 1457–1464.
29. Karatzas, D.; Shafait, F.; Uchida, S.; Iwamura, M.; Bigorda, L.G.I.; Mestre, S.R.; Mas, J.; Mota, D.F.; Almazan, A.; Heras, L.P.D.L. ICDAR 2013 robust reading competition. In Proceedings of the 2013 12th International Conference on Document Analysis and Recognition., Washington, WA, USA, 25–28 August 2013; pp. 1484–1493.
30. Lucas, S.M.; Panaretos, A.; Sosa, L. ICDAR 2003 robust reading competitions. In Proceedings of the Seventh International Conference on Document Analysis and Recognition, Edinburgh, UK, 6 August 2003; pp. 682–687.
31. Risnumawan, A.; Shivakumara, P.; Chan, C.S.; Tan, C.L. A robust arbitrary text detection system for natural scene images. *Expert Syst. Appl.* **2014**, *41*, 8027–8048. [CrossRef]

32. Mishra, A.; Alahari, K.; Jawahar, C.V. Scene text recognition using higher order language priors. In Proceedings of the British Machine Vision Conference 2012, Surrey, UK, 3–7 September 2012; pp. 1–11.
33. Rahman, M.A.; Wang, Y. Optimizing Intersection-Over-Union in deep neural networks for image segmentation. In *Advances in Visual Computing*; Springer International Publishing: Gewerbestrasse, Switzerland, 2016.
34. Bissacco, A.; Cummins, M.; Netzer, Y.; Neven, H. PhotoOCR: Reading Text in Uncontrolled Conditions. In Proceedings of the 2013 IEEE International Conference on Computer Vision (ICCV), Sydney Conference Centre, Darling Harbour, Sydney, 1–8 December 2013; pp. 785–792.
35. He, P.; Huang, W.L.; Qiao, Y.; Loy, C.C.; Tang, X.O.; Info, A. Reading scene text in deep convolutional sequences. In Proceedings of the National Conference on Artificial Intelligence, Phoenix, AZ, USA, 12–17 February 2016; pp. 3501–3508.
36. Jaderberg, M.; Simonyan, K.; Vedaldi, A.; Zisserman, A. Deep Structured Output Learning for Unconstrained Text Recognition. *Eprint Arxiv* **2014**, *24*, 603–611.
37. Lee, C.Y.; Osindero, S. Recursive recurrent nets with attention modeling for OCR in the wild. *Comput. Vis. Pattern Recognit.* **2016**, *43*, 2231–2239.
38. Shi, B.G.; Bai, X.; Yao, C. An End-to-End trainable neural network for Image-Based sequence recognition and its application to scene text recognition. *IEEE Trans. Pattern Anal. Mach. Intell.* **2016**, *39*, 2298–2304. [CrossRef] [PubMed]
39. Shi, B.G.; Wang, X.G.; Lyu, P.; Yao, C.; Bai, X. Robust scene text recognition with automatic rectification. *Comput. Vis. Pattern Recognit.* **2016**, *2*, 4168–4176.
40. Yang, X.; He, D.F.; Zhou, Z.H.; Kifer, D.; Giles, C.L. Learning to read irregular text with attention mechanisms. In Proceedings of the 26th International Joint Conference on Artificial Intelligence, Melbourne, Australia, 19–25 August 2017; pp. 3280–3286.

sensors

MDPI

Article

Pearson Correlation-Based Feature Selection for Document Classification Using Balanced Training

Inzamam Mashood Nasir [1], Muhammad Attique Khan [1], Mussarat Yasmin [2], Jamal Hussain Shah [2], Marcin Gabryel [3], Rafał Scherer [3] and Robertas Damaševičius [4,*]

1 Department of Computer Science, HITEC University, Taxila 47080, Pakistan; inzamam.mashood@hitecuni.edu.pk (I.M.N.); attique.khan@hitecuni.edu.pk (M.A.K.)
2 Department of Computer Science, COMSATS University Islamabad, Wah Campus, Wah Cantonment 47040, Pakistan; mussaratabdullah@gmail.com (M.Y.); jamalhussainshah@gmail.com (J.H.S.)
3 Department of Intelligent Computer Systems, Częstochowa University of Technology, 42-200 Częstochowa, Poland; marcin.gabryel@pcz.pl (M.G.); rafal.scherer@pcz.pl (R.S.)
4 Faculty of Applied Mathematics, Silesian University of Technology, 44-100 Gliwice, Poland
* Correspondence: robertas.damasevicius@polsl.pl

Received: 27 October 2020; Accepted: 25 November 2020; Published: 27 November 2020

Abstract: Documents are stored in a digital form across several organizations. Printing this amount of data and placing it into folders instead of storing digitally is against the practical, economical, and ecological perspective. An efficient way of retrieving data from digitally stored documents is also required. This article presents a real-time supervised learning technique for document classification based on deep convolutional neural network (DCNN), which aims to reduce the impact of adverse document image issues such as signatures, marks, logo, and handwritten notes. The proposed technique's major steps include data augmentation, feature extraction using pre-trained neural network models, feature fusion, and feature selection. We propose a novel data augmentation technique, which normalizes the imbalanced dataset using the secondary dataset RVL-CDIP. The DCNN features are extracted using the VGG19 and AlexNet networks. The extracted features are fused, and the fused feature vector is optimized by applying a Pearson correlation coefficient-based technique to select the optimized features while removing the redundant features. The proposed technique is tested on the Tobacco3482 dataset, which gives a classification accuracy of 93.1% using a cubic support vector machine classifier, proving the validity of the proposed technique.

Keywords: document classification; deep learning; feature selection; data augmentation; imbalanced dataset

1. Introduction

Document analysis and classification refer to automatically extracting the information and classifying it into a suitable category. Documents are often referred to as 2D material that can contain text or graphical items and can be used in optical character recognition (OCR) [1], word spotting [2], page segmentation [3], and cursive handwriting recognition [4] tasks. Document classification is considered as an essential step in classifying and analyzing the image documents. For several applications, classifying documents into their respective classes is a prerequisite step. If documents are well-sorted, it can be dispatched to the relative department for processing [5]. The indexing efficiency of a digital library can be improved with document classification [6]. Classifying the documents into content categories such as a table of content or a title page can suggest how pages extracting the metadata can be useful [7]. The retrieval efficiency and accuracy can be improved by classification on visual similarities, which can help users extract an article from any specific document or journal, containing a specific keyword, image, or table [8]. As the document classification is considered a higher-level analysis task, it is important to select the suitable document classes and types to get high accuracy and high performance in terms of effectiveness and efficiency [8].

Existing techniques either utilized the simple feedforward neural networks, standalone deep convolutional neural networks (DCNNs) models, or performed better on datasets, where a dataset contains a limited number of classes of documents. However, real-world cases have many issues in document classification, including structural similarities, low-quality images, and informational layers like signatures, marks, logo, and handwritten notes, which degrade the overall efficiency of many previously proposed methods. Data imbalance is also an essential problem in the deep learning (DL) domain, as overfitted and under fitted data can easily affect the overall performance of the proposed model. To resolve the problem of overfitting, the max-pooling layers have been added to the deep neural network models.

In this article, an automated system is proposed to classify the document images efficiently in accuracy and prediction time. We analyze and reduce the impact of adverse document image issues by employing multiple CNNs and combining each model's training and properties. The selected primary dataset Tobacco3482 is hugely imbalanced, which is tackled by proposing a novel data augmentation technique. The secondary dataset RVL-CDIP is used to populate the minority classes. The fusion of multiple networks produces redundant features, which are tuned by employing the Pearson correlation coefficient (PCC)-based optimization technique.

The structure of this article is as follows. Details of the proposed technique are described in Section 3. Experimental results to validate the proposed technique are presented in Section 4, and Section 5 concludes this article with a conclusion and future directions in this research field.

2. Literature Review

Classification based on the content of document images has been broadly contemplated. Document classification can be performed using the visual-based local document image [9]. Structure models like letters and forms gave interesting results, when classified using region-based algorithms [10]. Morphological features such as text skew and handwriting skew have been addressed using entropy algorithm [11] and projection profiling [12]. The study of documents is commonly dependent on text removed using OCR techniques [13]. In another case, OCR is inclined to errors and is not generally pertain to every type of documents, e.g., the handwritten content is yet hard to peruse. A 4-layer Convolutional Neural Network (CNN) model was utilized for document classification using a small tobacco dataset for classifying tax forms [14]. This experiment outperformed the previous Horizontal-Vertical Partitioning and Random Forest (HVP-RF) and Speeded Up Robust Features (SURF) descriptor-based classification technique achieving an accuracy of 65%. Another technique for document classification utilizes principal component analysis (PCA) along with one-class support vector machine (OCSVM) in which PCA reduced the dimensionality and OCSVM performed the classification [15]. The PCA initially chose the top features for the document images from four different datasets. Then OCSVM was trained on selected features to classify the images into the most relevant classes with a precision rate of 99.62%. A semi-supervised learning approach utilizing CNNs based on graph-structured data was presented in [16]. The main idea is to localize the convolutions in an approximation of first-order spectral graphs. The model initially scaled according to the number of graph edges. It started learning the representations of hidden layers that encoded the features on the nodes and structure of local graphs. The approach was demonstrated on three datasets having 6, 7, and 3 classes, respectively.

In another work, multi-label document classification is applied to Czech newspaper documents, where features are extracted using a simple multi-layer perceptron and convolutional networks [17]. The achieved F1 score for this method was 84.0% while using a multi-layer perceptron with sigmoid functions. A biomedical document classification was carried out in [18], where an imbalanced bio-dataset was used for a cluster-based classification on the under-sampled dataset GXD. Overall precision of 0.72 was achieved. Another method involving a region-based training for document classification was proposed, which utilized the properties of the VGG16 model via transfer learning and achieved an accuracy of 92.2% on the Ryerson Vision Lab Complex Document Information Processing (RVL-CDIP) dataset [19].

The recent success of CNN [20] is inspired by novel deep learning applications such as breast cancer classification [21], fashion product classification [22], text sentiment analysis [23], computer network security [24], medical image analysis for disease diagnostics [25], speech recognition [26], semantic segmentation [27], malware classification [28], remote sensing [29], and document image analysis [30]. The CNN process is known as a supervised learning method, in which features are extracted and classified by a learning algorithm. Compositions are performed on the learned vectors for classification using deep learning methods. The performance of these networks is improved by collecting larger datasets, learning more powerful models, and avoiding overfitting using better techniques. These larger datasets include ImageNet [31], consisting of more than 15 million labeled images in 22,000 different categories, and LabelMe [32], consisting of millions of fully segmented images. The CNNs can learn from the larger datasets using different models [33]. These models' capacity can be controlled by changing the order of layers to classify the input images correctly. As compared to the feedforward neural network with the same number of layers, CNNs contain fewer parameters and connections, making it easier and more convenient to test and train. As the use of graphical processing units (GPUs) has increased recently, many techniques have proposed effective and efficient ways to train CNNs using single and multiple GPUs [34]. After the success of a deep CNN model AlexNet [35], many more CNN models like GoogleNet [36], ZFNET [37], VGGNet [38], and ResNet [39] have also shown improved performance and results.

3. Proposed Method

For several applications, classifying documents into their respective classes is a prerequisite step. The indexing efficiency of a digital library can be enhanced with the help of document classification. There are numerous publicly accessible datasets for document classification, yet two acclaimed datasets, Tobacco3482 [40] and RVL-CDIP [41], are used, containing thousands of document images divided into 10 and 16 classes, respectively. These datasets have their challenges, and to get improved performance, a new technique utilizing the DCNN features is proposed having five significant steps, including (1) data balancing; (2) pre-processing; (3) feature extraction; (4) feature fusion, and (5) feature selection. In the first step, the imbalanced Tobacco3482 dataset is balanced using data augmentation technique. The dataset is then scaled down to the input sizes of both DCNN models and forwarded to pre-trained models, i.e., AlexNet and VGG19 to extract the DCNN features. Serial feature fusion is then applied on the DCNN features to fuse both models, which was finally optimized using the PCC-based technique [42]. These optimized features are forwarded to classifiers to obtain the classification accuracy. Additionally, a detailed model of the proposed technique is shown in Figure 1.

3.1. Data Augmentation

Imbalance of a dataset is a significant problem in any field as this can cause problems by ignoring the document images containing relevant information. Data imbalance occurs when one or more classes have a lower number of samples than the rest of the classes. Because of this problem, many well-modeled neural network architectures have failed to perform well. Imbalanced datasets in the domain of machine learning tend to produce unsatisfactory results. For any imbalanced dataset, if an event from minority class is predicted with an event rate of less than 5%, that is considered a rare event. The Logistic Regression and Decision Tree-based classification techniques tend to have a biased behavior toward rare events. These methods accurately predict the majority class, ignoring the minority class as noise. This eventually leaves a strong possibility of misclassifying the minority class when compared with the majority class.

This paper proposes a data augmentation-based approach to solve the data imbalance issue in an appropriate way. The following equations explain the process of solving this issue using the variables defined in Table 1.

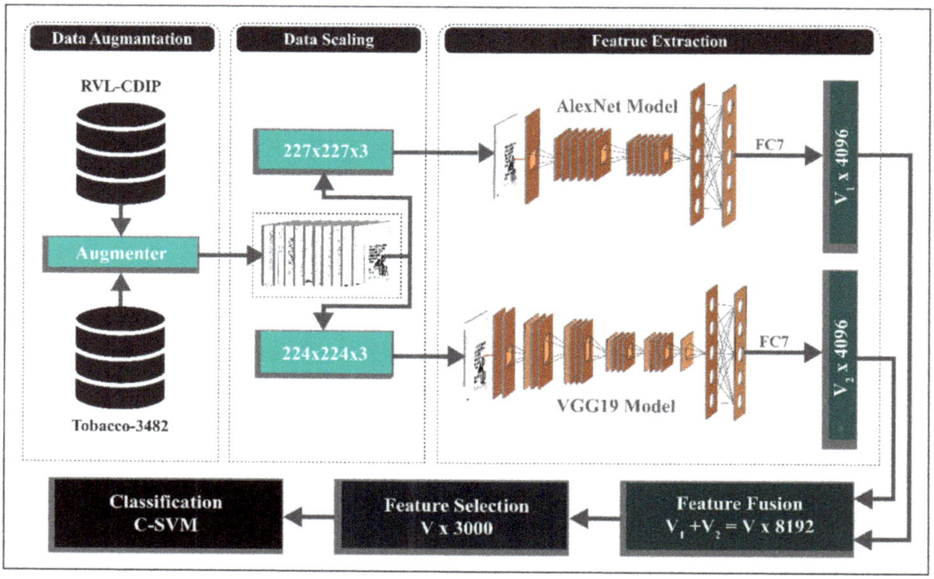

Figure 1. Detailed model of the proposed method.

Table 1. Nomenclature of variables used in the definitions and equations.

Variable	Description	Variable	Description
T	Threshold Value	C_i	Sum of images in the ith class
D	Difference between the threshold and the sum of a single class		
D_1	Tobacco3482 dataset	D_2	RVL-CDIP dataset
D_3	Balanced Dataset	In_b	Input from the previous neuron
Out_a	Output of the current neuron	$\omega_{a,b}$	Weight of the connection between ath and bth neuron
ξ	Activation function	V_1	DCNN features of AlexNet
V_2	DCNN features of VGG19	M_{FV_i}	The merit M of feature subset FV having i features
avg_{cf}	Feature-classification correlations	avg_{ff}	Feature-feature correlations
F_i	ith feature	W_i	ith weight

The threshold T is defined as following, which represents the highest class of the dataset:

$$T = \max(C_i), \tag{1}$$

where C_i represents the sum of images in the ith class and $i = 1, \ldots, n$.

$$D = \begin{cases} T - C_i, if\ C_i < T \\ 0,\ if\ C_i \geq T \end{cases}, \tag{2}$$

where D is the difference between the threshold and the sum of a single class, which is computed by comparing C_i with a threshold value. If D gives a non-zero value, it is forwarded to a function and the class label to fetch images from the secondary dataset to balance the primary dataset.

The flow diagram of the data augmenter is shown in Figure 2.

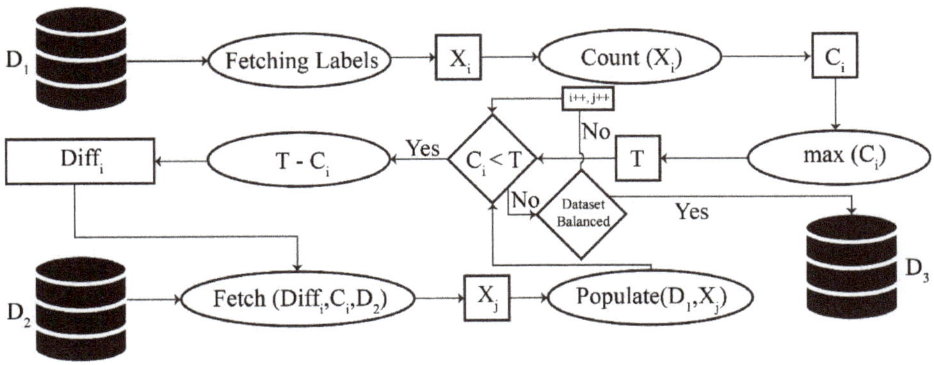

Figure 2. Flow diagram of data augmenter.

The algorithm for data balancing is mentioned below (see Algorithm 1). Here, the input is D_1 which denotes the Tobacco3482 dataset, while the output is D_3, which is an augmented, balanced dataset. Initially, all the labels are extracted from a dataset, which denotes all the classes. These labels are used to count images within each class, and a threshold value T is assigned with the highest-class count. The samples in all other classes are compared with T to calculate the difference. This difference, along with the class label and the secondary dataset D_2 is used to fetch the required number of images and populate the D_1 to form a new augmented dataset D_3.

Algorithm 1. Dataset balancing using a secondary dataset

Input: D_1
Output: D_3
Step 1: $X_i \leftarrow D_1$
Step 2: $C_i \leftarrow Count(X_i),\ where\ i = 1, \ldots, n$
Step 3: $T \leftarrow \max(C_i)$
Step 4: $Diff_i \leftarrow \begin{cases} T - C_i,\ if\ C_i < T \\ 0 \quad ,\ if\ C_i \geq T \end{cases}$
Step 5: $X_j \leftarrow Fetch(Diff_i, C_i, D_2)$
Step 6: $D_3 = Populate\left(D_1, X_j\right)$
return D_3

The comparison of the primary dataset before and after augmentation is shown in Table 2. The classes in the primary and secondary datasets are also inserted in the table to make the comparison understandable. RVL-CDIP is a secondary dataset to balance the primary dataset (Tobacco3482). Table 2 shows the classes of both datasets. Left-most column present class names in a primary dataset, while the right-most column presents the corresponding classes from the RVL-CDIP dataset. The central columns present the number of images before and after data augmentation.

Table 2. Dataset before and after applying the data augmentation algorithm.

Classes in Tobacco3482	# of Images before Augmentation	# of Images after Augmentation	Classes in RVL-CDIP
Advertisement	230	620	Advertisement
Email	599	620	Email
Form	431	620	Form
Letter	567	620	Letter
Memo	620	620	Memo
News	188	620	News Article
Note	201	620	Handwritten
Report	265	620	Scientific Report
Resume	180	620	Resume
Scientific	261	620	Scientific Publication

3.2. Network Architectures

Transfer of information between neurons is the primary motivation of CNNs. The CNNs have the same basic structure as classical artificial networks. The CNNs are composed of multiple layers which continuously fire neurons among connecting layers. The previous layer fires neurons onto the next layer as input, and each of these connections of successive layers is burdened with values called weights. The major difference between CNNs and classical networks is that classical networks accept the inputs in the form of vectors, while CNNs accept images as input data. The convolutional layer is the first layer of CNN, which receives an image from the input layer, and it uses an operation called image convolution to extract the features. To understand the functionality, a filter $f_{m,n}$ of size 3×3 is defined with a central position at m, n.

Many CNN models have pooling layers with each convolutional layer, which reduces the input image by selecting fewer pixels based on three major operations known as "max-pooling" "min-pooling", and "average-pooling". A pooling filter of size 3×3 will select only one value, which replaces all the nine values in the new vector representing the input image. The last layers of CNN models are always fully connected layers and separated into output layers or hidden layers. A tiny image described by numerical values is the input to these layers, which is already rectified by the previous combinations of convolutional and pooling layers. This layer uses an activation function to extract features from the rectified input image by creating multiple neurons and identifying the total units with each pixel value. The working of neurons can be described as:

$$Out_a = \xi(\sum_b^n \omega_{a,b} In_b), \tag{3}$$

where Out_a is an output of the current neuron, In_b is input from the previous neuron, $\omega_{a,b}$ is the weight of the connection between ath and bth neuron and ξ is the activation function which is used to normalize the input values received from previous neurons to the range of $(-1, 1)$ can be further described as:

$$\xi(In) = tanh(In), \tag{4}$$

3.2.1. AlexNet

The AlexNet has eight (8) distinguished layers, out of which five connected convolutional layers are at the beginning with pooling layers, followed by three (3) fully-connected layers. The output layer of this model is the softmax layer, which is directly connected with the last fully connected layer. The last layer is labeled as the FC8 layer, which fed the softmax layer with a feature vector of 1000 size, and softmax produces 1000 channels. Neurons of fully connected layers are directly attached to neurons of previous layers. Normalization layers relate to first and second layers. Fifth convolutional layer and response normalization layers have max-pooling layers. The output of every fully connected and convolutional layer has a ReLU layer. Input size for this network is $227 \times 227 \times 3$. The AlexNet model structure used in this technique is shown in Figure 3 where FC7 is selected as an output layer.

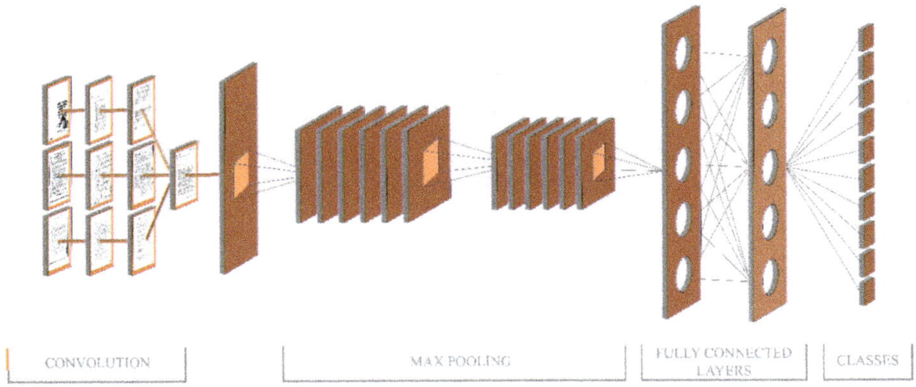

Figure 3. Structure of AlexNet Model.

3.2.2. VGG19

Depth is an essential aspect of the CNN architecture. Increasing the layers of the network by adding more layers, a more significant CNN architecture was developed, which was more accurate on the ImageNet Large Scale Visual Recognition Challenge (ILSVRC) classification and localization tasks. The input to the VGG19 architecture is a fixed size RBG image of $224 \times 224 \times 3$. Multiple convolutional layers accept the input image, which has the smallest sized 3×3 filters. The 1×1 convolutional filter was also used to transform the input channel from non-linearity to linear. One-pixel convolution stride is fixed, and the spatial resolution is fixed by the spatial padding for the convolutional layer. Five max-pooling layers carry the spatial pooling, out of which convolutional layers follow few. Having stride of 2, over a 2×2 pixel window, maximum pooling is applied. VGG19 also has three fully connected layers followed by a softmax layer at the end. The structure of the VGG19 model is explained in the following Figure 4, where FC7 is an output layer.

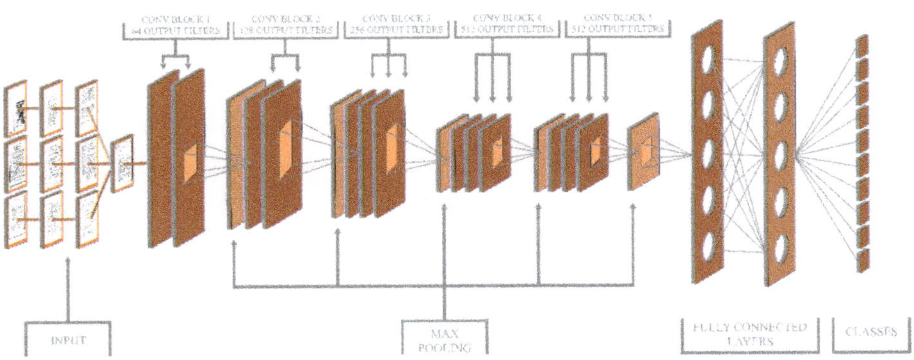

Figure 4. Structure of VGG19 Model.

3.3. *Feature Fusion and Selection*

After extracting the deep features using two DCNN networks, AlexNet and VGG19, both features are serially fused to form a higher dimensional feature vector, which is explained as follows.

Suppose $a_1, a_2, a_3, \ldots, a_n$ belongs to a feature space V_1 and $b_1, b_2, b_3, \ldots, b_n$ belongs to the feature space V_2, and feature spaces V_1 and V_2 denote the DCNN features of AlexNet and VGG19, respectively. Feature spaces V_1 and V_2 are defined as:

$$V_1 = \begin{bmatrix} a_{1,1} & a_{1,2} & \cdots & a_{1,4096} \\ a_{2,1} & a_{2,2} & \cdots & a_{2,4094} \\ \vdots & \vdots & \vdots & \vdots \\ a_{n,1} & a_{n,2} & \cdots & a_{n,4096} \end{bmatrix}, \tag{5}$$

$$V_2 = \begin{bmatrix} b_{1,1} & b_{1,2} & \cdots & b_{1,4096} \\ b_{2,1} & b_{2,2} & \cdots & b_{2,4094} \\ \vdots & \vdots & \vdots & \vdots \\ b_{n,1} & b_{n,2} & \cdots & b_{n,4096} \end{bmatrix}, \tag{6}$$

$$FV = V_1 \oplus V_2, \tag{7}$$

$$FV = \begin{bmatrix} a_{1,1} & a_{1,2} & \cdots & a_{1,4096} & b_{1,4097} & b_{1,4098} & \cdots & b_{1,8192} \\ a_{2,1} & a_{2,2} & \cdots & a_{2,4096} & b_{2,4097} & b_{2,4098} & \cdots & b_{2,8192} \\ \vdots & \vdots & \vdots & \vdots & \vdots & \vdots & \vdots & \vdots \\ a_{n,1} & a_{n,2} & \cdots & a_{n,4096} & b_{n,4097} & b_{n,4098} & \cdots & b_{n,8192} \end{bmatrix}, \tag{8}$$

where FV is a fused feature vector.

As both networks were trained to extract the features from fully connected layer FC7, a total of 4096 features were extracted and fused to form a new feature vector of size 8192 features. This fusion process compensates the inadequacy of a single network for document classification but increases the feature vector's dimensions. Moreover, both networks use a basic CNN architecture with different approaches; there are chances of many correlations and redundant features among fused features.

Therefore, in this work, a PCC-based technique is implemented for selecting the optimized features by removing the redundant ones. The PCC-based feature selection technique evaluates different subsets of features based on highly correlated features [43].

The following equation explains the merit M of feature subset FV having i features:

$$M_{FV_i} = \frac{i \times avg_{cf}}{\sqrt{i + i(i-1)avg_{ff}}}, \tag{9}$$

where avg_{cf} corresponds to the feature-classification correlations while avg_{ff} corresponds to feature-feature correlations.

The criterion for the correlation coefficient-based feature selection $CCFS$ can be defined as:

$$CCFS = \max_{FV_i} \left[\frac{avg_{cf_1} + avg_{cf_2} + \ldots + avg_{cf_i}}{\sqrt{i + 2(avg_{f_1 f_k} + \ldots + avg_{f_m f_n} + \ldots + avg_{f_i f_1})}} \right], \tag{10}$$

where avg_{cf_i} and $avg_{f_m f_n}$ are referred to as correlations between continuous features.

Suppose W_i denotes the whole feature vector having F_i features, then the equation mentioned above for $CCFS$ can be rewritten as an optimized feature vector as:

$$CCFS = \max_{w \in \{0,1\}^i} \left[\frac{\left(\sum_{j=1,k=1}^{i} f_i w_i \right)^2}{\sum_{j=1}^{i} w_i + \sum_{j \neq k} 2 \times f_j w_j w_{ij}} \right], \tag{11}$$

Features having a high correlation value are considered as redundant features, so only those features are selected, which have the minimum redundancy between consecutive features. The smallest Pearson's correlation values concerning neighboring features are appended to the selected feature set. The feature vector's final size becomes 3000 after selecting the best features and disregarding the redundant features. These best features are forwarded to the Cubic SVM (C-SVM) classifier to obtain the classification accuracy. The proposed technique is tested on the publicly available dataset Tobacco3482. The labeled outputs of the proposed technique are shown in Figure 5.

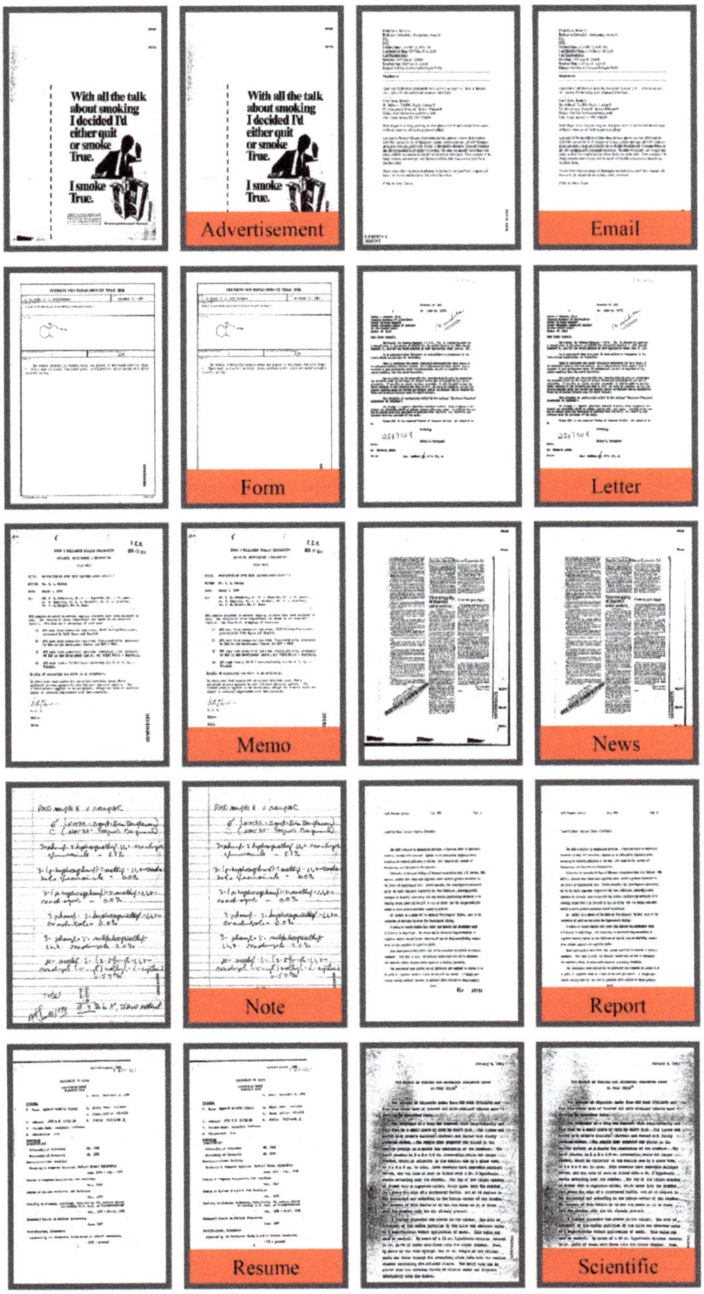

Figure 5. Labeled outputs of the proposed technique.

4. Experimental Results

4.1. Datasets

The publicly available Tobacco3482 dataset is presented by a tobacco company including a different number of pictures per class, having 3482 pictures of high resolution from ten different classes. These images have a remarkable difference in structural and visual views, making this dataset more complex and challenging. The RVL-CDIP dataset is also a complicated, huge dataset that includes 400,000 labeled images in 16 different categories. In this article, RVL-CDIP was used as a secondary dataset for the augmentation purpose. The proposed technique is validated on the original Tobacco3482 dataset and an augmented dataset prepared during the data augmentation process. Few sample images from the Tobacco3482 dataset are shown in Figure 6.

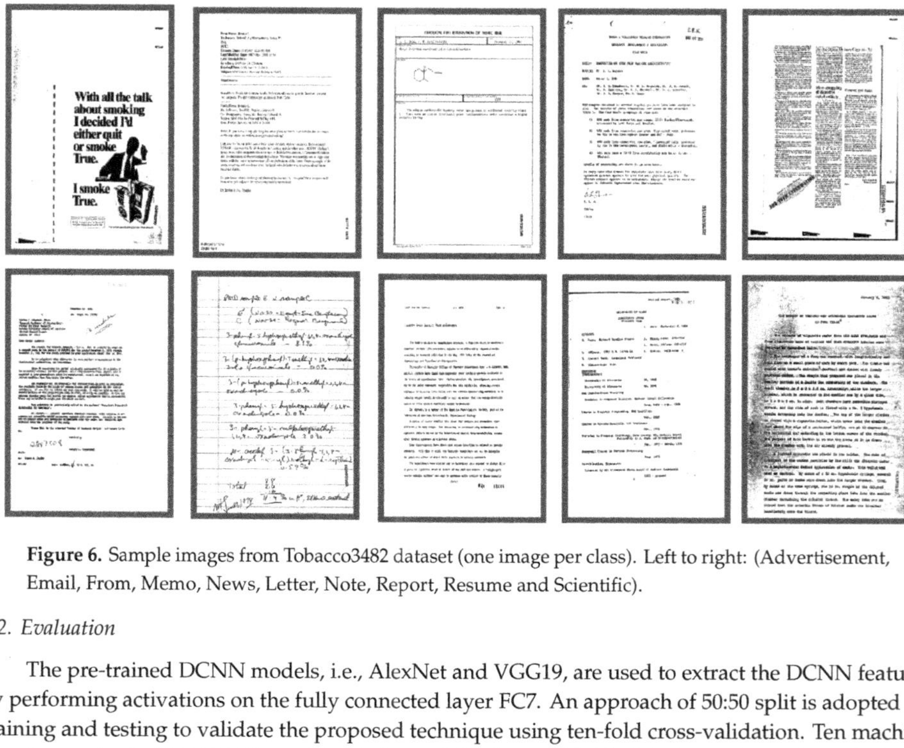

Figure 6. Sample images from Tobacco3482 dataset (one image per class). Left to right: (Advertisement, Email, From, Memo, News, Letter, Note, Report, Resume and Scientific).

4.2. Evaluation

The pre-trained DCNN models, i.e., AlexNet and VGG19, are used to extract the DCNN features by performing activations on the fully connected layer FC7. An approach of 50:50 split is adopted for training and testing to validate the proposed technique using ten-fold cross-validation. Ten machine learning methods (C-SVM, Linear Discriminant (LD), linear SVM (L-SVM), quadratic SVM (Q-SVM), fuzzy KNN (F-KNN), modified KNN (M-KNN), continuous KNN (C-KNN), weighted KNN (W-KNN), Subspace Discriminant, and Subspace KNN) were used as classifiers. All experiments are performed on Corei7, 7th generation with a 3.4 GHz processor, 16 GB RAM, 256 GB SSD having MATLAB 2018a (MathWorks Inc., Natick, MA, USA).

4.3. Classification Results

Three experiments are performed to obtain classification results such as (a) classification using the AlexNet features with PCC-based optimization; (b) classification using VGG19 features with the PCC-based optimization; (c) classification using a fusion of AlexNet and VGG19 features with the PCC-based optimization. Classification accuracy and execution time are validated by comparing it with the state-of-the-art techniques applied to the same dataset and sub-dataset.

AlexNet DCNN with PCC-based Optimization: In the first experiment, the AlexNet model is used to extract DCNN features that are reduced using the PCC-based optimization to select the best features. Selected 3000 features were then forwarded to ten (10) different classifiers. The best classification accuracy of 90.1% and false-negative rate (FNR) of 9.9% is achieved using C-SVM with a training time of 670.8 s. The confusion matrix, shown in Figure 7a, confirms the accuracy of C-SVM. Q-SVM achieves the second-best accuracy with 89.6% and FNR of 10.4% in execution time of 742.2 s. Overall results of this experiment on different classifiers is displayed in Table 3.

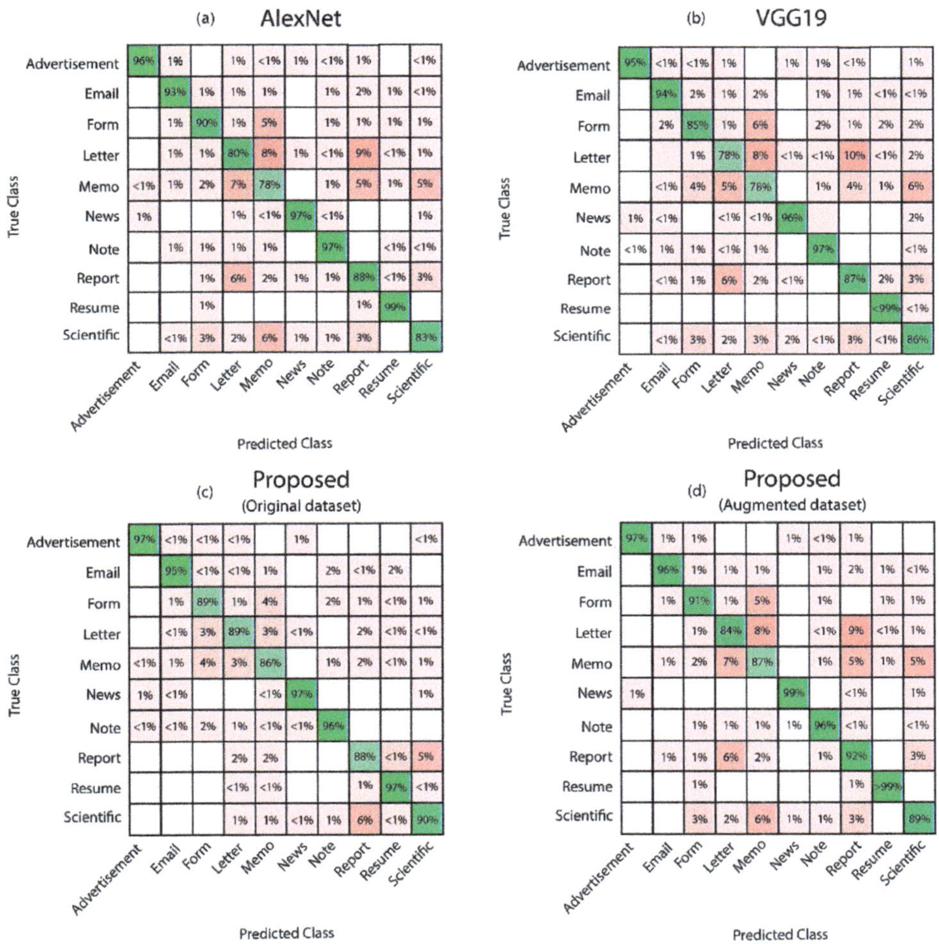

Figure 7. Confusion matrices for the Tobacco3482 dataset: (**a**) AlexNet, (**b**) VGG19, (**c**) proposed method on the original dataset, and (**d**) proposed method on the augmented dataset.

Table 3. Comparison of classification accuracy, false-negative rate (FNR), and Training Time on Tobacco3482 Dataset. Best values are shown in bold.

	Experiments				Performance Measures		
Method	**AlexNet**	**VGG-19**	**Proposed (Original Dataset)**	**Proposed (Augmented Dataset)**	**Accuracy (%)**	**FNR (%)**	**Training Time (s)**
C-SVM	√				**90.1**	**9.0**	**670.8**
		√			**89.6**	**10.4**	**947.3**
			√		**92.2**	**7.8**	**329.5**
				√	**93.1**	**6.9**	**364.1**
Linear Discriminant	√				79.7	20.3	772.5
		√			-	-	-
			√		82.4	17.6	593.2
				√	84.0	16.0	659.9
L-SVM	√				81.6	18.4	1731.7
		√			79.0	21.0	2198.9
			√		81.8	18.2	971.3
				√	84.3	15.7	1170.2
Q-SVM	√				89.6	10.4	742.2
		√			87.1	12.9	1996.0
			√		87.4	12.6	582.6
				√	91.8	8.2	625.2
F-KNN	√				87.1	12.9	846.8
		√			83.7	16.3	1720.9
			√		85.0	15.0	742.6
				√	89.5	10.5	872.9
M-KNN	√				73.8	26.2	744.9
		√			65.5	34.5	1920.3
			√		73.9	26.1	621.1
				√	76.5	23.5	767.4
C-KNN	√				74.0	26.0	1604.8
		√			65.9	34.1	4147.5
			√		72.8	27.2	598.4
				√	76.4	23.6	719.1
W-KNN	√				87.1	12.9	951.4
		√			83.0	17.0	2393.5
			√		84.3	15.7	687.2
				√	88.7	11.3	746.9
Subspace Discriminant	√				89.5	10.3	5305.0
		√			87.5	12.5	6304.3
			√		88.3	11.7	1716.8
				√	89.7	10.3	2079.2
Subspace KNN	√				87.0	13.0	2498.6
		√			83.2	16.8	2508.9
			√		86.9	13.1	1958.2
				√	89.4	10.6	2398.9

VGG19 DCNN with PCC-based Optimization: In this experiment, VGG19 is used for DCNN feature extraction and PCC selected the optimized features. Selected 3000 features are then forwarded to ten (10) different classifiers, out of which, the best classification accuracy at 89.6% and FNR of 10.4% is recorded on C-SVM with a training time of 947.3 s. The classification accuracy of Cubic SVM is confirmed by the confusion matrix shown in Figure 7b. The second highest accuracy of 87.1% with FNR of 12.9%, and training time of 1996 s was achieved on Q-SVM. The detailed results of this experiment on multiple classifiers are listed in Table 3 as well.

AlexNet and VGG19 DCNN feature fusion and PCC-based Optimization: A serial-based fusion approach is applied to fuse the DCNN features of AlexNet and VGG19 models, which are later optimized using the PCC-based selection. Both DCNN models extracted 4096 features each, and feature fusion strategy is applied to combine both models' characteristics.

The proposed technique is validated on two cases for a fair comparison with existing techniques. Initially, the proposed technique is validated using the original imbalanced Tobacco3482 dataset, where it achieved the highest accuracy of 92.2% with FNR of 7.8% and training time of 329.5 s on C-SVM classifier. While in another case, it is validated using an augmented dataset after the augmentation process described in the proposed section, where the original dataset was balanced using a secondary dataset RVL-CDIP. C-SVM achieved the best accuracy of 93.1% in 364.1 s with FNR of 6.9%. Figure 7c,d shows the confusion matrices, which confirms classification accuracy of Cubic SVM on both cases. Table 4 contains the results of all experiments mentioned above on ten selected classifiers along with respective accuracies, FNR, and training time. There are other experiments, which are carried out to validate the proposed model. Table 4 illustrates the results after feature fusion. The highest accuracy of 91.5% is achieved using C-SVM. It is noteworthy that this experiment's training time increases as the total number of features increased after fusion. The fusion increases the chances of redundant and irrelevant features, which are removed by employing PCC-based feature selection technique.

Table 4. Classification results after feature fusion. Best values are shown in bold.

Classifier	Performance Measures					
	Sensitivity (%)	Precision (%)	AuC (%)	FNR (%)	Accuracy (%)	Training Time (s)
C-SVM	**91.6**	**91.6**	**99.3**	**8.50**	**91.5**	3037.7
Linear Discriminant	81.2	81.3	89.7	18.7	81.3	3055.7
L-SVM	84.5	84.8	98.1	15.6	84.4	2861.1
Q-SVM	90.3	90.3	99.0	9.70	90.3	2989.4
F-KNN	86.7	87.0	92.6	13.2	86.8	2176.9
M-KNN	75.0	76.3	95.5	24.9	75.1	2176.7
C-KNN	74.5	76.0	95.3	25.5	74.5	5307.6
W-KNN	86.9	87.4	98.5	13.0	87.0	**2174.4**
Subspace Discriminant	86.1	86.2	98.5	13.7	86.3	8794.6
Subspace KNN	86.9	87.0	95.5	13.0	87.0	3876.5

4.4. Discussion

We discuss the significance of proposed results on several classifiers. Without statistical analysis, it is not clear that which classifier outperforms for document classification. Therefore, we have conducted more experiments and computed standard deviation, confidence interval (CI), denoted by $\sigma_{\bar{x}}$ and margin of error at confidence level (95%, 1.96 $\sigma_{\bar{x}}$). The values are tabulated in Tables 5 and 6. In Table 5, the minimum accuracy achieved on C-SVM after 100 iterations is 90.7% whereas the average and best accuracies are 91.45% and 92.2%, respectively. The value of σ is 0.75 and $\sigma_{\bar{x}}$ is 0.5303, respectively. The margin of error on confidence level (CL) (95%, 1.96 $\sigma_{\bar{x}}$) is 91.45 ± 1.039 (±1.14%), which is better as compared to other classifiers. Similarly, the analysis is also conducted on the augmented dataset and values are tabulated in Table 6. For C-SVM, CL (95%, 1.96 $\sigma_{\bar{x}}$) is 92.7 ± 0.554 (±0.60%), which is better as compared to other classifiers performance.

Table 5. Analysis of proposed method on original data. Best values are shown in bold.

Method	Min (%)	Avg (%)	Max (%)	σ	$\sigma_{\bar{x}}$	ME (95%, 1.96 $\sigma_{\bar{x}}$)
C-SVM	**90.7**	**91.45**	**92.2**	**0.75**	**0.5303**	**91.45 ± 1.039 (±1.14%)**
LD	79.4	80.90	82.4	1.5	1.0606	80.9 ± 2.079 (±2.57%)
L-SVM	78.3	80.05	81.8	1.75	1.2374	80.05 ± 2.425 (±3.03%)
Q-SVM	84.8	86.10	87.4	1.3	0.9192	86.1 ± 1.802 (±2.09%)
F-KNN	83.2	84.10	85.0	0.9	0.6363	84.1 ± 1.247 (±1.48%)
M-KNN	70.6	72.25	73.9	1.65	1.6670	72.25 ± 2.87 (±3.17%)
C-KNN	71.1	71.95	72.8	0.85	0.6010	71.95 ± 1.178 (±1.64%)
W-KNN	81.6	82.95	84.3	1.35	0.9545	82.95 ± 1.871 (±2.26%)
ESDA	85.4	86.85	88.3	1.45	1.0253	86.85 ± 2.010 (±2.31%)
ESKNN	83.2	85.05	86.9	1.85	1.3081	85.05 ± 2.564 (±3.01%)

Table 6. Analysis of proposed method on augmented dataset. Best values are shown in bold.

Method	Min (%)	Avg (%)	Max (%)	σ	$\sigma_{\bar{x}}$	ME (95%, 1.96 $\sigma_{\bar{x}}$)
C-SVM	**92.3**	**92.7**	**93.1**	**0.4**	**0.2828**	**92.7 ± 0.554 (±0.60%)**
LD	81.7	82.8	84.0	1.15	0.8131	82.85 ± 1.594 (±1.92%)
L-SVM	82.6	83.4	84.3	0.85	0.6010	83.45 ± 1.178 (±1.41%)
Q-SVM	89.4	90.6	91.8	1.2	0.8485	90.6 ± 1.663 (±1.84%)
F-KNN	87.1	88.3	89.5	1.2	0.8485	88.3 ± 1.663 (±1.88%)
M-KNN	73.8	75.1	76.5	1.35	0.9545	75.1 ± 1.871 (±2.59%)
C-KNN	73.6	75.1	76.4	1.4	0.9899	75.0 ± 1.940 (±2.59%)
W-KNN	84.9	86.8	88.7	1.9	1.3435	86.8 ± 2.633 (±3.03%)
ESDA	85.7	87.7	89.7	2.0	1.4142	87.7 ± 2.772 (±3.16%)
ESKNN	86.3	87.8	89.4	1.55	1.0969	87.85 ± 2.148 (±2.45%)

Several previous techniques had also used the Tobacco3482 dataset to validate their models. A custom CNN-based architecture, inspired by AlexNet, was proposed in [44] for document classification. Multiple experiments were performed including 20 images per class and 100 images per class for training and validation, respectively, and achieved classification accuracies of 68.25% and 77.6%, for both tests respectively. Another approach utilized DCNN model as a feature extractor and extreme learning machine (ELM) for classification in [45]. Overall accuracy of 83.24% was achieved on the Tobacco3482 dataset. A DCNN-based approach utilizing AlexNet, VGG16, GoogLeNet, and ResNet-50 was proposed in [46], where classification accuracy of 91.13% is recorded. In [47], a spatial pyramid model is proposed to extract high discriminant multi-scale features of document images by utilizing the inherited layouts of images. A deep multi-column CNN model is used to classify the images with an overall classification accuracy of 82.78%. In [48], combining semantic information with visual information of images allowed an improved separation toward document classification. The model has tested on the Tobacco800 [49] dataset and achieved an accuracy of 93%. Tobacco-800 is a subset of the actual Tobacco3482 dataset, with fewer classes. The purpose of comparing this dataset is to validate the proposed methodology demonstrating that it still outperforms other techniques tested with less classes. The performance of related work is summarized in Table 7.

Table 7. Comparison with existing techniques on the Tobacco3482 dataset.

Paper	Dataset	Accuracy (%)	Training Time (s)	Prediction Time (s)
Afzal et al. [44]	Tobacco3482	77.6	-	-
Kölsch et al. [45]	Tobacco3482	83.24	-	-
Afzal et al. [46]	Tobacco3482	91.13	-	-
Sarkhel & Nandi [47]	Tobacco3482	82.78	-	-
Wiedemann & Heyer [48]	Tobacco-800	93	-	-
Proposed	Primary:	AlexNet: 90.1	670.8	2.34
	Tobacco3482	VGG19: 89.6	947.3	3.95
	Secondary:	Original: 92.2	329.5	1.62
	RVL-CDIP	Augmented: 93.1	364.1	0.78

The proposed technique obtained a classification accuracy of 93.1% with an average training time of 364.17 s and an average prediction time of 0.78 s. Note that the proposed technique's training time increases when it is tested on the augmented dataset due to the increased number of images in each class. But as the training proceeds, the prediction time is reduced in half, which shows the balanced dataset's importance.

5. Conclusions

In this article, a hybrid approach to classify the documents using deep convolutional neural networks is proposed, consisting of data augmentation, data normalization, feature extraction, feature fusion, and feature selection steps. In the data augmentation step, the dataset is analyzed, and classes within the dataset with fewer images are fed using the secondary dataset RVL-CDIP. After that, data normalization is performed, which resized the dataset images according to pre-trained models' sizes. The pre-trained AlexNet and VGG19 models are used to extract deep features, which are fused using a serial-based fusion, and, in the end, the Pearson correlation coefficient-based technique selects the best features. The selected features are then forwarded to the Cubic SVM classifier for document classification. The proposed technique is validated on the publicly available Tobacco3482 dataset, achieving an accuracy of 93.1%. The obtained results outperformed the previous techniques and validated the proposed technique.

Moreover, this technique reduces training and prediction time, which is also an essential development in the document classification field. There are several open questions for this research including: (a) The selection of CNN models (other pre-trained or custom CNN models may perform better on this domain); (b) the selection of the technique to fuse different features is also not a limitation, as there are several other fusion techniques [50–53], which can perform better; and (c) feature selection technique used in this work is also not a limitation as other feature selection methods can also be implemented and tested.

In the future, a new generic method for document image classification will be developed by combining the hand-crafted features with the DCNN features to achieve a further improved classification accuracy. Furthermore, a real-time application also will be developed to classify documents in real-time.

Author Contributions: Conceptualization, M.A.K. and J.H.S.; methodology, M.A.K. and M.Y.; software, I.M.N.; validation, M.A.K. and R.D.; formal analysis, M.G., R.S. and R.D.; investigation, I.M.N., M.A.K., M.Y. and J.H.S.; resources, I.M.N. and M.Y.; data curation, I.M.N.; writing—original draft preparation, I.M.N., M.A.K., M.Y. and J.H.S.; writing—review and editing, M.G., R.S. and R.D.; visualization, I.M.N. and M.A.K.; supervision, M.A.K.; project administration, R.S.; funding acquisition, R.S. All authors have read and agreed to the published version of the manuscript.

Funding: This research received no external funding.

Conflicts of Interest: The authors declare no conflict of interest.

References

1. He, S.; Schomaker, L. Beyond OCR: Multi-faceted understanding of handwritten document characteristics. *Pattern Recognit.* **2017**, *63*, 321–333. [CrossRef]
2. Giotis, A.P.; Sfikas, G.; Gatos, B.; Nikou, C. A survey of document image word spotting techniques. *Pattern Recognit.* **2017**, *68*, 310–332. [CrossRef]
3. Chen, K.; Seuret, M.; Liwicki, M.; Hennebert, J.; Ingold, R. Page segmentation of historical document images with convolutional autoencoders. In Proceedings of the 13th International Conference on Document Analysis and Recognition (ICDAR), Tunis, Tunisia, 23–26 August 2015; pp. 1011–1015. [CrossRef]
4. Samanta, O.; Roy, A.; Parui, S.K.; Bhattacharya, U. An HMM framework based on spherical-linear features for online cursive handwriting recognition. *Inf. Sci.* **2018**, *441*, 133–151. [CrossRef]
5. Noce, L.; Gallo, I.; Zamberletti, A. Query and Product Suggestion for Price Comparison Search Engines based on Query-product Click-through Bipartite Graphs. In Proceedings of the 12th International Conference on Web Information Systems and Technologies, WEBIST 2016, Rome, Italy, 23–25 April 2016; Volume 1, pp. 17–24.
6. Crowe, J.P. Library Indexing System and Method. U.S. Patent US20150066945A1, 1 January 2019.
7. Zamberletti, A.; Noce, L.; Gallo, I. Text localization based on fast feature pyramids and multi-resolution maximally stable extremal regions. In Proceedings of the Asian Conference on Computer Vision, ACCV 2014, Singapore, 1–5 November 2014; pp. 91–105. [CrossRef]
8. Gallo, I.; Zamberletti, A.; Noce, L. Interactive object class segmentation for mobile devices. In Proceedings of the 27th SIBGRAPI Conference on Graphics, Patterns and Images (SIBGRAPI), Rio de Janeiro, Brazil, 26–30 August 2014; pp. 73–79. [CrossRef]
9. Sauvola, J.; Pietikäinen, M. Adaptive document image binarization. *Pattern Recognit.* **2000**, *33*, 225–236. [CrossRef]
10. Hu, J.; Kashi, R.; Wilfong, G. Comparison and classification of documents based on layout similarity. *Inf. Retr.* **2000**, *2*, 227–243. [CrossRef]
11. Brodić, D.; Milivojević, Z.N. Text skew detection using combined entropy algorithm. *Inf. Technol. Control* **2017**, *46*, 308–318. [CrossRef]
12. Ptak, R.; Żygadło, B.; Unold, O. Projection-based text line segmentation with a variable threshold. *Int. J. Appl. Math. Comput. Sci.* **2017**, *27*, 195–206. [CrossRef]
13. Akhtar, Z.; Lee, J.W.; Khan, M.A.; Sharif, M.; Khan, S.A.; Riaz, N. Optical character recognition (OCR) using partial least square (PLS) based feature reduction: An application to artificial intelligence for biometric identification. *J. Enterp. Inf. Manag.* **2020**. [CrossRef]
14. Tensmeyer, C.; Martinez, T. Analysis of convolutional neural networks for document image classification. In Proceedings of the 14th IAPR International Conference on Document Analysis and Recognition (ICDAR), Kyoto, Japan, 9–15 November 2017; pp. 388–393. [CrossRef]
15. Kumar, B.S.; Ravi, V. Text Document Classification with PCA and One-Class SVM. In Proceedings of the 5th International Conference on Frontiers in Intelligent Computing: Theory and Applications, FICTA 2016, Bhubaneswar, India, 16–17 September 2016; pp. 107–115. [CrossRef]
16. Kipf, T.N.; Welling, M. Semi-supervised classification with graph convolutional networks. In Proceedings of the 5th International Conference on Learning Representations, ICLR 2017, Toulon, France, 24–26 April 2017.
17. Lenc, L.; Král, P. Deep neural networks for Czech multi-label document classification. In Proceedings of the International Conference on Intelligent Text Processing and Computational Linguistics, Konya, Turkey, 3–9 April 2016; pp. 460–471.
18. Jiang, X.; Ringwald, M.; Blake, J.A.; Arighi, C.; Zhang, G.; Shatkay, H. An effective biomedical document classification scheme in support of biocuration: Addressing class imbalance. *Database* **2019**, *2019*. [CrossRef]
19. Das, A.; Roy, S.; Bhattacharya, U.; Parui, S.K. Document Image Classification with Intra-Domain Transfer Learning and Stacked Generalization of Deep Convolutional Neural Networks. In Proceedings of the 24th International Conference on Pattern Recognition (ICPR), Beijing, China, 20–24 August 2018; pp. 3180–3185. [CrossRef]
20. Rashid, M.; Khan, M.A.; Sharif, M.; Raza, M.; Sarfraz, M.M.; Afza, F. Object detection and classification: A joint selection and fusion strategy of deep convolutional neural network and SIFT point features. *Multimed. Tools Appl.* **2019**, *78*, 15751–15777. [CrossRef]

21. Nasir, I.M.; Rashid, M.; Shah, J.H.; Sharif, M.; Awan, M.Y.H.; Alkinani, M.H. An Optimized Approach for Breast Cancer Classification for Histopathological Images Based on Hybrid Feature Set. *Curr. Med. Imaging* **2020**, *16*. [CrossRef] [PubMed]
22. Nasir, I.-M.; Khan, M.A.; Alhaisoni, M.; Saba, T.; Rehman, A.; Iqbal, T.A. Hybrid Deep Learning Architecture for the Classification of Superhero Fashion Products: An Application for Medical-Tech Classification. *Comput. Model. Eng. Sci.* **2020**, *124*, 1–9. [CrossRef]
23. Kapočiūtė-Dzikienė, J.; Damaševičius, R.; Woźniak, M. Sentiment analysis of lithuanian texts using traditional and deep learning approaches. *Computers* **2019**, *8*, 4. [CrossRef]
24. Wei, W.; Ke, Q.; Nowak, J.; Korytkowski, M.; Scherer, R.; Woźniak, M. Accurate and fast URL phishing detector: A convolutional neural network approach. *Comput. Netw.* **2020**, *178*, 107275. [CrossRef]
25. Khan, M.A.; Ashraf, I.; Alhaisoni, M.; Damaševičius, R.; Scherer, R.; Rehman, A.; Bukhari, S.A.C. Multimodal brain tumor classification using deep learning and robust feature selection: A machine learning application for radiologists. *Diagnostics* **2020**, *10*, 565. [CrossRef] [PubMed]
26. Pipiras, L.; Maskeliūnas, R.; Damaševičius, R. Lithuanian Speech Recognition Using Purely Phonetic Deep Learning. *Computers* **2019**, *8*, 76. [CrossRef]
27. Zhang, M.; Jing, W.; Lin, J.; Fang, N.; Wei, W.; Woźniak, M.; Damaševičius, R. NAS-HRIS: Automatic design and architecture search of neural network for semantic segmentation in remote sensing images. *Sensors* **2020**, *20*, 5292. [CrossRef]
28. Nisa, M.; Shah, J.H.; Kanwal, S.; Raza, M.; Khan, M.A.; Damaševičius, R.; Blažauskas, T. Hybrid malware classification method using segmentation-based fractal texture analysis and deep convolution neural network features. *Appl. Sci.* **2020**, *10*, 4966. [CrossRef]
29. Sun, Z.; Lin, D.; Wei, W.; Woźniak, M.; Damasevicius, R. Road detection based on shearlet for GF-3 synthetic aperture radar images. *IEEE Access* **2020**, *8*, 28133–28141. [CrossRef]
30. Bella, F.Z.A.; El Rhabi, M.; Hakim, A.; Laghrib, A. Reduction of the non-uniform illumination using nonlocal variational models for document image analysis. *J. Frankl. Inst.* **2018**, *355*, 8225–8244. [CrossRef]
31. Deng, J.; Dong, W.; Socher, R.; Li, L.-J.; Li, K.; Fei-Fei, L. Imagenet: A large-scale hierarchical image database. In Proceedings of the IEEE Conference on Computer Vision and Pattern Recognition, CVPR 2009, Miami, FL, USA, 20–25 June 2009. [CrossRef]
32. Russell, B.C.; Torralba, A.; Murphy, K.P.; Freeman, W.T. LabelMe: A database and web-based tool for image annotation. *Int. J. Comput. Vis.* **2008**, *77*, 157–173. [CrossRef]
33. Alom, M.Z.; Taha, T.M.; Yakopcic, C.; Westberg, S.; Sidike, P.; Nasrin, M.S.; Hasan, M.; Van Essen, B.C.; Awwal, A.A.S.; Asari, V.K. A State-of-the-Art Survey on Deep Learning Theory and Architectures. *Electronics* **2019**, *8*, 292. [CrossRef]
34. Li, X.; Zhang, G.; Huang, H.H.; Wang, Z.; Zheng, W. Performance analysis of GPU-based convolutional neural networks. In Proceedings of the International Conference on Parallel Processing, Philadelphia, PA, USA, 16–19 August 2016; pp. 67–76. [CrossRef]
35. Krizhevsky, A.; Sutskever, I.; Hinton, G.E. Imagenet classification with deep convolutional neural networks. *Commun. ACM* **2017**, *60*, 84–90. [CrossRef]
36. Szegedy, C.; Liu, W.; Jia, Y.; Sermanet, P.; Reed, S.; Anguelov, D.; Rabinovich, A. Going deeper with convolutions. In Proceedings of the IEEE Conference on Computer Vision and Pattern Recognition, Boston, MA, USA, 8–10 June 2015; pp. 1–9.
37. Zeiler, M.D.; Fergus, R. Visualizing and understanding convolutional networks. In Proceedings of the European Conference on Computer Vision, Zurich, Switzerland, 6–12 September 2014; pp. 818–833. [CrossRef]
38. Simonyan, K.; Zisserman, A. Very deep convolutional networks for large-scale image recognition. *arXiv* **2014**, arXiv:1409.1556.
39. He, K.; Zhang, X.; Ren, S.; Sun, J. Deep residual learning for image recognition. In Proceedings of the IEEE Conference on Computer Vision and Pattern Recognition, Las Vegas, NV, USA, 27–30 June 2016; pp. 770–778. [CrossRef]
40. Blum, A. On-line algorithms in machine learning. In *Online Algorithms;* Fiat, A., Woeginger, G.J., Eds.; Springer: Berlin/Heidelberg, Germany, 1998; pp. 306–325.
41. Harley, A.W.; Ufkes, A.; Derpanis, K.G. Evaluation of Deep Convolutional Nets for Document Image Classification and Retrieval. In Proceedings of the 13th International Conference on Document Analysis and Recognition ICDAR 2015, Tunis, Tunisia, 23–26 August 2015; pp. 991–995. [CrossRef]

42. Stigler, S.M. Francis Galton's account of the invention of correlation. *Stat. Sci.* **1989**, *4*, 73–79. [CrossRef]
43. Senliol, B.; Gulgezen, G.; Yu, L.; Cataltepe, Z. Fast Correlation Based Filter (FCBF) with a different search strategy. In Proceedings of the 23rd International Symposium on Computer and Information Sciences, ISCIS'08, Istanbul, Turkey, 27–29 October 2008; pp. 1–4. [CrossRef]
44. Afzal, M.Z.; Capobianco, S.; Malik, M.I.; Marinai, S.; Breuel, T.M.; Dengel, A.; Liwicki, M. Deepdocclassifier: Document classification with deep convolutional neural network. In Proceedings of the 13th International Conference on Document Analysis and Recognition (ICDAR), Tunis, Tunisia, 23–26 August 2015; pp. 1111–1115. [CrossRef]
45. Kölsch, A.; Afzal, M.Z.; Ebbecke, M.; Liwicki, M. Real-time document image classification using deep CNN and extreme learning machines. In Proceedings of the 14th IAPR International Conference on Document Analysis and Recognition (ICDAR), Kyoto, Japan, 9–15 November 2017. [CrossRef]
46. Afzal, M.Z.; Kölsch, A.; Ahmed, S.; Liwicki, M. Cutting the error by half: Investigation of very deep cnn and advanced training strategies for document image classification. In Proceedings of the 14th IAPR International Conference on Document Analysis and Recognition (ICDAR), Kyoto, Japan, 9–15 November 2017. [CrossRef]
47. Sarkhel, R.; Nandi, A. Deterministic routing between layout abstractions for multi-scale classification of visually rich documents. In Proceedings of the 28th International Joint Conference on Artificial Intelligence, Macao, China, 10–16 August 2019; pp. 3360–3366.
48. Wiedemann, G.; Heyer, G. Multi-modal page stream segmentation with convolutional neural networks. *Lang. Resour. Eval.* **2019**, 1–24. [CrossRef]
49. Lewis, D.; Agam, G.; Argamon, S.; Frieder, O.; Grossman, D.; Heard, J. Building a test collection for complex document information processing. In Proceedings of the 29th Annual Int. ACM SIGIR Conference (SIGIR 2006), Seattle, WA, USA, 6–11 August 2006; pp. 665–666.
50. Arshad, H.; Khan, M.A.; Sharif, M.I.; Yasmin, M.; Tavares, J.M.R.S.; Zhang, Y.-D.; Satapathy, S.C. A multilevel paradigm for deep convolutional neural network features selection with an application to human gait recognition. *Expert Syst.* **2020**, e12541. [CrossRef]
51. Khan, M.A.; Zhang, Y.-D.; Khan, S.A.; Attique, M.; Rehman, A.; Seo, S. A resource conscious human action recognition framework using 26-layered deep convolutional neural network. *Multimed. Tools Appl.* **2020**, 1–23. [CrossRef]
52. Khan, M.A.; Sharif, M.I.; Raza, M.; Anjum, A.; Saba, T.; Shad, S.A. Skin lesion segmentation and classification: A unified framework of deep neural network features fusion and selection. *Expert Syst.* **2019**, e12497. [CrossRef]
53. Khan, M.A.; Akram, T.; Sharif, M.; Javed, K.; Rashid, M.; Bukhari, S.A.A. An integrated framework of skin lesion detection and recognition through saliency method and optimal deep neural network features selection. *Neural Comput. Appl.* **2020**, *32*, 15929–15948. [CrossRef]

sensors

:le

xt Detection Using Multi-Stage Region Proposal Network nsitive to Text Scale †

nito Nagaoka, Tomo Miyazaki *, Yoshihiro Sugaya and Shinichiro Omachi

Graduate School of Engineering, Tohoku University, Sendai 9808579, Japan; naga.yoshi.yoshi@gmail.com (Y.N.); sugaya@iic.ecei.tohoku.ac.jp (Y.S.); machi@ecei.tohoku.ac.jp (S.O.)
* Correspondence: tomo@tohoku.ac.jp
† This paper is an extended version of our paper published in Nagaoka, Y.; Miyazaki, T.; Sugaya, Y.; Omachi, S. Text Detection by Faster R-CNN with Multiple Region Proposal Networks. In Proceedings of the 7th International Workshop on Camera-Based Document Analysis and Recognition (CBDAR), Kyoto, Japan, 9–15 November 2017; pp. 15–20.

Abstract: Recently, attention has surged concerning intelligent sensors using text detection. However, there are challenges in detecting small texts. To solve this problem, we propose a novel text detection CNN (convolutional neural network) architecture sensitive to text scale. We extract multi-resolution feature maps in multi-stage convolution layers that have been employed to prevent losing information and maintain the feature size. In addition, we developed the CNN considering the receptive field size to generate proposal stages. The experimental results show the importance of the receptive field size.

Keywords: scene text detection; multiple scales; convolutional neural networks

tion: Nagaoka, Y.; Miyazaki, T.; aya, Y.; Omachi, S. Text Detection g Multi-Stage Region Proposal work Sensitive to Text Scale †. sors 2021, 21, 1232. https:// org/10.3390/s21041232

demic Editor: Kyandoghere makya

eived: 29 December 2020
epted: 5 February 2021
lished: 9 February 2021

blisher's Note: MDPI stays neu- with regard to jurisdictional clai- in published maps and institutio- affiliations.

1. Introduction

Recently, attention has surged concerning intelligent sensors using text detection [1,2]. Texts in a natural scene image are useful for many applications, such as translator, mobile visual search, and so on. Thus, text detection is a hot topic in computer vision. A convolutional neural network, CNN, is widely used in object detection tasks since its high performance. Particularly, Faster R-CNN [3] is a standard method. Moreover, there are YOLO [4–6] and SSD [7]. Text detection benefits from CNN-based object detection to achieve high performance.

It is unsuitable to directly apply object detection methods [3–7] to text detection. As shown in Figure 1a, the small texts "reuse" and "in" were missed in the left example, and large text "lowns-uk.co" was divided in the right example. The CNNs transformed images into a low-resolution feature maps. Thus, some texts are transformed to appropriate scales in the feature maps. However, small and large texts became inappropriate scales, resulting in detection failures.

There is room in Faster R-CNN to improve scale sensitivity. Its limited scale sensitivity is due to a fixed receptive field in region proposal network (RPN), a Faster R-CNN module. RPN extracts context features around objects using convolutional computation. The receptive field size of the RPN is essential. The convolutional computation produces one pixel in a feature map from a fixed area context. For example, 3×3 kernel of the convolution produces one output from 3×3 input. The receptive field depends on the number of convolutional computations. In the case of Faster R-CNN, the receptive field is 228×228. We doubt whether Faster R-CNN can utilize the context information well because it has a fixed receptive field.

The problem of small object detection is caused by detection from only one feature map. Recently, multi-stage convolutional feature maps [8,9] are applied to many works for not only object detection but also other tasks. While this strategy is useful, but there are few discussions about quantitative analysis. He et al. [10] introduced a skip-connection

module to prevent overfitting, which was the first attempt to merge different feature m
Wang et al. [11] explained the effectiveness of using convolutional layers simultaneo
These explain the effectiveness of using multi-feature maps; however, there are no deta
works on the receptive field, to our knowledge. Our proposed idea computes the recep
field size. Therefore, it can extract adequate context features for generating propo
Besides, the proposed idea can be applied to other detection modules and tasks.

(**a**) Faster R-CNN

(**b**) The proposed method

Figure 1. Detection examples. (**a**) Faster R-CNN (convolutional neural network) failed to detect sm texts and detected large texts only partially. Green circles are the missed texts. (**b**) The propos method detected small and large texts successfully. Although, there is the false-positive detection the left example.

To reinforce the scale sensitivity, we propose a CNN that can detect small and lar texts simultaneously. Specifically, we propose to use multiple RPNs to generates text pr posals in different resolution feature maps. These multiple RPNs have different recepti field sizes. As shown in Figure 1b, the proposed method detected small and large tex successfully. The contribution of this paper is the integration of Faster R-CNN and a mul resolution detection approach using multiple anchors of the appropriate dimensions f texts. The proposed architecture is sensitive to text region scale by using a multi-recepti field size. We confirm that the receptive field is an important factor when using the CN and the proposed concept can contribute to other detection methods.

This paper is an extended version of our conference paper [12]. There are fo differences from the conference paper. Firstly, we reorganized the related work sectic using more than 20 additional literature to clarify the background of the proposed metho Secondly, we conducted an ablation study to confirm improvements of the two propose components, multiple RPNs and Anchor. Section 4.3 summarizes the results. Thirdly, w visualized the output of each RPN to confirm output scales are appropriate. Section 4 showed that text detection is performed by RPNs that are responsible for small and larg scale, respectively. Finally, we analyzed failure results by investigating the output of th RPNs and activated feature maps. Section 4.5 illustrated the output. Overall, these fou additional discussions and experiments reinforced the conference paper.

2. Related Works

A text detection method is based on object detection. Hence, we describe object detection methods. Then, we address some studies to use multi-resolution feature maps for object detection. Finally, we introduce text detection studies.

2.1. Object Detection

Object detection is a popular research subject in computer vision. There have been many attempts, such as deformable part model [13] and histograms of sparse codes [14] which use engineered feature expression and support vector machine. These methods incur high computation cost because they need many feature expressions and parameters for evaluation. Recently, the CNN-based method and R-CNN [15] have been used for object detection. R-CNN is composed of a proposal generation stage and a classification stage. Proposals from a given image are generated using modules of other methods such as Selective Search [16]. The proposal regions cropped from an original input image are fed into the classification stage, which uses the CNN to classify proposals into the object or background classes. In addition, the bounding-box regression process adjusts the proposal rectangles to object sizes accurately. The problem of R-CNN is high computation cost because the CNN computes the feature map for each proposal. In the Fast R-CNN [17], RoI-pooling (region of interest pooling) is introduced to share precomputed convolution features. Given an input image, the CNN computes the feature maps of the whole image. The feature maps of the proposal regions are cropped and pooled to the fixed size by using RoI-pooling. This reduces the computation cost; however, the Fast R-CNN requires another pipeline to generate proposals. Therefore, it cannot process end-to-end consistently. The Faster R-CNN [3] uses the RPN to generate proposals with only convolutional layers. In the RPN, the convolutional layer (3×3 kernel) is applied to obtain the feature map, which is fed into two sibling convolutional layers (1×1 kernel) for binary classification (object/background) and bounding-box regression. In each pixel position of the feature map, some proposals with confidence scores are generated from fixed-size rectangles called anchors in the bounding-box regression. Therefore, the Faster R-CNN does not require an external proposal generating method by RPN module. The Faster R-CNN is a baseline method for achieving state-of-the-art accuracy and inference speed. This realizes end-to-end processing and improves the detection speed and accuracy.

YOLO (you only look once) [4–6] is a one-shot detector and is not a region-based method. It predicts proposals with object likelihood scores and class probabilities. Therefore, it does not need any computation modules per proposal. This leads to less computation than the Faster R-CNN. SSD (single shot multibox detector) [7] is similar to YOLO, except for using a multi-resolution feature map for detection. SSD predicts the proposals from each convolutional layer. Therefore, it has various features for detection, unlike the Faster R-CNN and YOLO.

2.2. Strategy Using Multi-Features

The CNN is composed of many convolutional layers, e.g., 13 layers in VGG16 [18]. In general, a shallow layer extracts simple features of an image, called as a low-level feature, and a deeper layer can extract complex features, called as a high-level feature. Therefore, many works using the CNN use many convolutional layers. However, using many convolutional layers incurs high computation cost. To avoid this, a downsampling operation called pooling is inserted after some convolutional blocks; however, it leads to loss of feature information as a trade-off. Many recent works have pointed out this phenomenon, particularly in object detection, face detection, and text detection.

A recent trend of using multi-stage convolutional feature maps is called feature pyramid. Kong et al. [9] pointed out that region-based methods struggle with small-size objects. To solve this problem, they use conv1, conv3, and conv5 feature maps of VGG16 and merge them into one feature map. This generates large-size feature maps using multi-feature states. Kong et al. [19] merge the convolutional feature map and deeper feature for accurate object localization. Lin et al. [8] applied feature merging to the Faster R-C and concluded that using feature hierarchy saves memory cost. Wang et al. [11] us multi-convolutional layer for high-order statistics to represent feature maps with neglig computation cost.

These strategies are inspired by skip-connection [10], and it leads to semantic segn tation [20–22] along with detection. In this work, we also considered receptive fields of multi-stage convolutional layer.

2.3. Text Detection

Text detection has been widely studied for decades. Wang et al. [23] detected c acters using the sliding window and random ferns [24] and connected the charact using pictorial structures [25]. Wang et al. [26] detected word regions using the slid window and CNN, and recognized characters using the CNN and dictionary-matchi Milyaev et al. [27] binarized images and generated word proposals integrated from c nected components by edge information and engineered features such as position a color. The character proposals classified by AdaBoost were connected to word prop als, which were followed by recognizing the word proposals by OCR (optical charac recognition). Opitz et al. [28] generated a text region confidence map using the s ing window and AdaBoost, and detected word regions by maximally stable extren region [29]. After detection, they recognized the text using CNN from a pre-defined dic nary. Jaderberg et al. [30] used edge boxes [31] and aggregate channel features detector [to generate text proposals and eliminate false positive proposals using random forest. Th they used the CNN for bounding-box regression and recognizing characters. Tian et al. [generated character proposals using the sliding window and fast cascade boosting al rithm [34] and connected the characters using the CNN. These methods involve multi-sta processing and complex pipeline. Hence, they require fine parameter tuning for generati proposals and classifying them. Recently, the deep learning approach has been frequen used because it does not require engineered features. In addition to this, the CNN-bas detection approach involves a simple architecture, realizing end-to-end consistent flc without complexity.

Therefore, many approaches are based on the recent progress in the end-to-end proce of object detection. Liao et al. [35] proposed end-to-end CNN-based SSD, employing a ho zontally long anchor to detect the text region efficiently. SSD uses multi-stage convolutior feature maps. Therefore, this approach is close to our proposed method. Tian et al. [3 predicted parts of the text region using the RPN to predict vertically long proposals ha ing fixed widths. The proposals are connected by bi-directional LSTM (long short ter memory), and the final output is the bounding-boxes of the text regions. Zhong et al. [3 improved the Faster R-CNN for text detection. By introducing an inception module [3 they used convolutional operations having multi receptive field and this leads to extra features efficiently compared with the conventional convolutional layer.

Recently, segmentation-based approaches are often employed. Tang et al. [39] use three CNNs for text region segmentation: One predicts the text region roughly, the secor one refines the text region pixels, and the last one judges whether the text region is correct not. Dai et al. [40] combined the Faster R-CNN and segmentation for arbitrary-oriented te This predicts the text mask after generating the proposals. Lyu et al. [41] predicted positio sensitive segmentation, which is robust to arbitrarily inclined text positions. Zhou et al. [4 proposed segmentation- and parameterize-inclined text region by expressing the distanc from the pixels. Bounding-boxes were generated based on the distance from one pixel i the text mask. This approach has simple architecture and can predict arbitrary coordinate of the bounding-boxes. He et al. [43] also predicted the parameters of the relative position for the bounding-boxes using segmentation strategy with fully convolutional network.

Not only text detection but also recognition methods are studied for recognizing word by CRNN (convolutional recurrent neural networks) [44] using connectionist temporal cla sification loss [45]. Bušta et al. [46] predicted the text region using an anchor-based detecto such as Faster R-CNN, and each region was recognized using the CRNN. Li et al. [47] combined the LSTM with the Faster R-CNN to realize text spotting (detection and recognition). First, the Faster R-CNN block outputs text bounding-boxes, and the two LSTMs, encoder LSTM and decoder LSTM, recognize the word in the bounding-box. This method detects text and recognizes end-to-end consistently using one deep learning model. Liu et al. [48] also combined text detection and recognition. In the text detection stage, this predicts arbitrarily oriented regions such as [42]. In the recognition stage, the proposals are rotated by affine transformation and are inputted in the CRNN module containing bi-directional LSTM and outputs labels.

Thus, the text detection methods have progressed notably in the virtue of CNN. We applied Faster R-CNN for object detection because this can be expanded to many works and be used as a baseline.

3. The Proposed Methods

In this section, we describe the proposed CNN module and its core concept.

3.1. *Scale-Sensitive Pyramid*

The proposed architecture is depicted in Figure 2. The main difference between Faster R-CNN and the proposed method is the total number of RPNs. While Faster R-CNN has one RPN in conv5-3 of VGG16, the proposed method has four RPNs in each convolutional layer. Specifically, RPN1, 2, 3, and 4 are added to conv4-6, conv5-3, conv6-3, and conv7-3, respectively. To use a large receptive field in the proposed architecture, we added two convolutional blocks containing one max-pooling and three convolutional layers such as VGG16. In addition, we used deep-feature representation in the conv4 stage and added extra three convolutional layers after conv4-3.

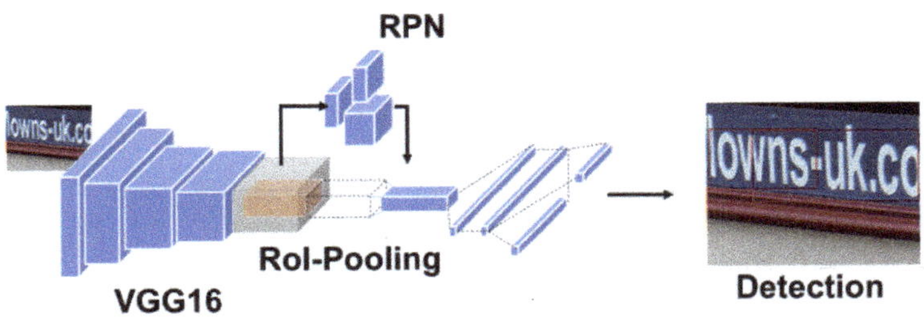

(**a**) Faster R-CNN

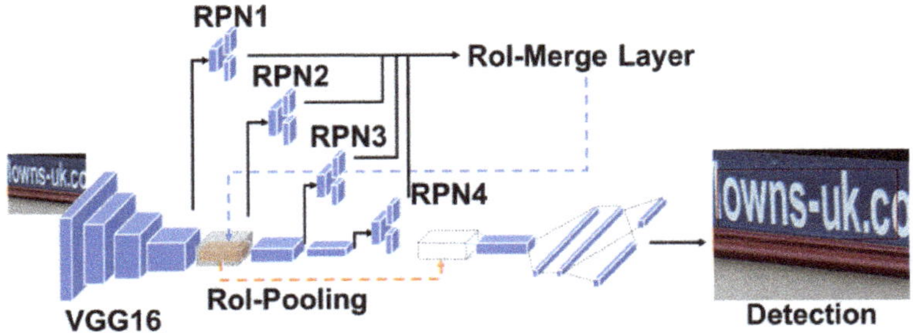

(**b**) The proposed method

Figure 2. Architectures.

In this paper, we define the number of RPNs as four to consider two purposes. Firstly, we aim to maximize two evaluation metrics, Recall and Precision. There is a trade-off between them. We can obtain a higher recall value by increasing the number of RPNs s more RPNs produce more text candidates. In contrast, the precision value decreases as number of candidates increases. Thus, we determined the number of RPNs heuristically considering the trade-off. Secondly, we aim to make training stable and feasible. There be more training parameters when the number of RPNs increased. Consequently, trair will be unstable. Besides, the amount of GPU memory is limited. Therefore, fou a feasible amount of RPNs for training. Although there is no experimental supp the above purposes are based on general facts. The trade-off between recall and precis is widely known. Moreover, training may be unstable if learning parameters increa Thus, we believe the reasons are convincing.

The RPNs generate proposals using each pixel of the feature maps. Thus, the propo were largely influenced by the convolutional layers. The convolutional layer having 3 kernel gathers 3 × 3 the size context in the input feature map to one pixel as the out Therefore, two accumulated convolutional layers gather 5 × 5 the context to one pi Considering this for an input image, we can determine the context size in the input ima which influences the generation of proposals in the RPN. In this paper, we denote t context size as a receptive field. The RPN of Faster R-CNN has a 228 × 228 size of receptive field. However, it is not sufficient to obtain information for detection, consider that the input size is about 600 × 600. On the other hand, the proposed method has fo RPNs, which have various receptive fields. The receptive fields of the RPN1, 2, 3, an are 156 × 156, 228 × 228, 468 × 468, 948 × 948, respectively. Therefore, while RPN1 can fine context to generate tiny proposals, RPN3 and RPN4 can use a large context to encl large text. We call this proposed architecture SSP-RPNs (scale-sensitive pyramid RP for convenience.

The SSP-RPNs have more RPNs than Faster R-CNN. Therefore, we introduce an R merge layer to prevent the increase in the computation cost for the proposals. The RoI-me layer receives 400 proposals (each RPN outputs 100 proposals) and applies non-maxim suppression to eliminate the overlapped proposals. Then, it selects up to 100 proposals a higher confidence score as output proposals.

3.2. Anchor for Text Detection

Anchor is rectangular with a fixed size in the RPN, and this is regressed to arbitra sized nonlinear transformation called bounding-box regression. However, transformati parameters are determined from anchor's height or width. Hence, the proposals are mai dependent on the anchor. Thus, we need to select an efficient anchor size for text detectio The main target of this work is Latin scripts containing alphabets and digits, and we c consider Latin scripts to be horizontally long instead of vertically long.

First of all, we performed the statistics for the text sizes in natural scene imag Figure 3 shows that the histogram result of the aspect ratio (width/height) in three trai ing datasets: COCO Text [49], Synth Text [50], and our dataset described in Section 4 The reliability of the histogram is based on diversity in the datasets. Specifically, our datas is composed of five public datasets, which are widely used in text detection studi Furthermore, COCO Text and Synth Text are large datasets containing 173 K texts and 8 words, respectively. The histograms shows that the text bounding-boxes are horizontal long, and particularly the half of them have widths two to four times the height. Fast R-CNN prepares various anchors depicted in Figure 4a. It contains horizontally long an vertically long aspect ratios of 1:2, 1:1, 2:1. Considering the statistics of the text boundin boxes, a vertically long anchor is unnecessary, and we need more horizontally long ancho In addition to this, the demand for a small-scale anchor increases because of the smalle receptive field size of the SSP-RPNs module of 156 × 156.

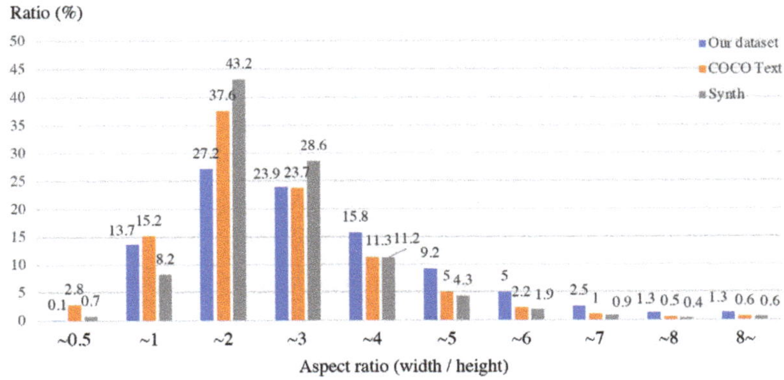

Figure 3. Histogram of the aspect ratio of texts in Datasets.

Based on the above reasons, we proposed new anchors for text detection shown in Figure 4b. We eliminated vertically long anchors and added horizontally long anchors for the Latin text. Moreover, we added a small-scale anchor for tiny text. For large-scale text, Faster R-CNN prepares large scale anchors, and we do not add any large-scale anchors. In the experiments, we confirmed that the proposed anchors were more efficient than the default anchors, and the anchor was an important factor for generating proposals.

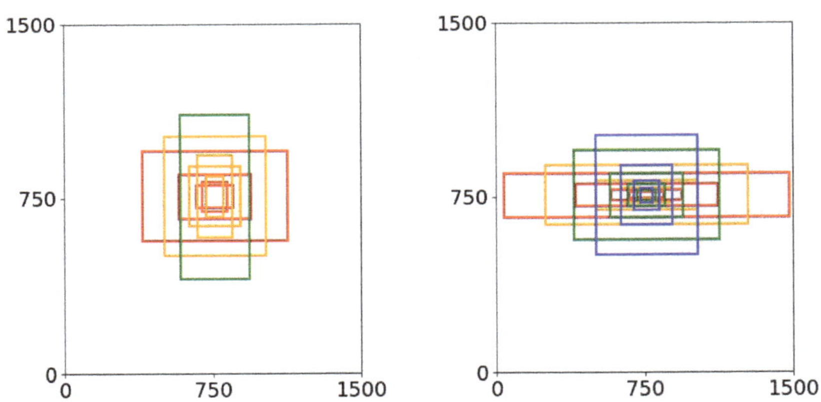

(**a**) Faster R-CNN (**b**) The proposed method

Figure 4. Comparison on anchors.

3.3. *Training Strategy*

The total loss for the proposed method is Equation (1).

$$L_{total} = \sum_{i \in \{1,2,3,4\}} \lambda_i L_{rpn_i} + \lambda_{fastrcnn} L_{fastrcnn} \tag{1}$$

L_{rpn_i} represents the loss of each RPN, and $L_{fastrcnn}$ is the loss of Fast R-CNN. λ_* means the hyper parameter to define the loss balance, we set $\lambda_* = 1$ in the experiments. L_{rpn_i} and $L_{fastrcnn}$ are composed of the classification loss and bounding-box regression loss, respectively. Detailed explanation can be found in [3,17].

We assign ground-truths to RPNs according to their sizes. Let ground-truth's size be maximum of either height or width. RPN1 is responsible for sizes less than 140. Followed by [3], RPN2 undertakes all ground-truths. Both of RPN3 and RPN4 take responsibility for sizes larger than 220. Overall, RPN1 is trained to be suitable for small-scale text, RPN3 RPN4 are used for large-scale text.

4. Experiments

In this section, we evaluated the proposed method and compared it with other detection methods. In training, the proposed model's parameters were initialized us ImageNet pretrained model, and the layers other than VGG16 were initialized accord to Gaussian distribution (mean is 0, the standard deviation is 0.01). The learning r was fixed to 0.001, weight decay was 0.0005, momentum was 0.9, and we iterated 10 For both training and testing, we used GPU NVIDIA TITAN X (Pascal). We implemen the proposed method using the faster R-CNN based on the deep learning framework, C (Implementation of Faster R-CNN with Caffe: https://github.com/rbgirshick/py-fas rcnn accessed on 29 December 2020).

4.1. Datasets and Evaluation Metrics

We compiled our training dataset including 7152 natural scene images contain texts. Our dataset is composed of five public datasets: ICDAR2013 RRC focused sce text training dataset (229 images) [51], ICDAR2015 RRC incidental scene text traini dataset (1000 images) [51], ICDAR2017 RRC multi lingual text training dataset (5425 i ages) [52], street view text training dataset (SVT Dataset: http://vision.ucsd.edu/~kai/s accessed on 29 December 2020), and KAIST dataset (KAIST Dataset: http://www.iapr-1.org/mediawiki/index.php/KAIST_Scene_Text_Database accessed on 29 December 20 (398 images). We evaluated the methods on the ICDAR2013 RRC focused scene text t dataset (233 images).

We used DetEval [53] containing three evaluation protocols, recall, precisi and F-score. The Recall represents that how much ground-truth is covered by the detecti results. Precision means that how accurately the methods generate the bounding-box F-score is the harmonic mean between recall and precision.

4.2. Numerical Results

The numerical results are shown in Table 1. The full results are available onli (Online results (Proposed): https://rrc.cvc.uab.es/?ch=2&com=evaluation&view=metho info&task=1&m=50094 accessed on 29 December 2020)

We compared the proposed method to other methods [3,33,35,37,43,50]. Particular Faster R-CNN [3] is an essential baseline of the proposed method. The fundament difference is the number of RPNs: one in the Faster R-CNN, four in the proposed metho Using only one RPN struggles with detecting small and large texts. Therefore, we propos to use four RPNs that are responsible for small and large texts, respectively. To verify t effectiveness of using four RPNs, a comparison with Faster R-CNN is necessary.

The proposed method outperformed Faster R-CNN more than seven points at F-sco Thus, we confirmed that the scale sensitivity could bring a certain improvement to te detection. Moreover, we showed the results of the proposed method in competition mod The full results are available online (Online results (Proposed, Competition mode): http //rrc.cvc.uab.es/?ch=2&com=evaluation&view=method_info&task=1&m=51720 accesse on 29 December 2020).

The comparison methods can be divided into two approaches in the aspect of sca strategy: multi-scale [35,43,50] and single-scale [3,33,37]. The multi-scale approach pr duces multiple resolution images using various scale ratios. A post-processing is require to merge results in multiple images. The single-scale approach uses a single resolutio image and applies multiple-sized kernels to detect various scaled texts.

According to the numerical results, the multi-scale methods were superior to th single-scale methods. Especially, the results of [43] are better because of the number c input images, such as seven images by scale ratios, $2^{\{-5,\cdots,1\}}$. The abundant input image are essential in the multi-scale approach. However, simultaneous detection for small an large texts is difficult in the multi-scale approach since small texts are collapsed easily. On the other hand, the proposed method keeps small and large texts intact. The multiple RPNs search texts in different resolution feature maps extracted from only one single image. As shown in Figure 5, the proposed method can detect various texts containing tiny-scale and large-scale texts.

Table 1. Numerical results on ICDAR2013.

Method	Input Scale	Recall	Precision	F-Score	Time
Gupta+. [50]	Multiple	75.5	92.0	83.0	-
He+ [43]	Multiple	81	92	86	0.9 s
Liao+. [35]	Multiple	83	89	86	0.73 s
Tian+ [33]	Single	75.9	85.2	80.3	-
Zhong+ [37]	Single	83	87	85	1.7 s
Baseline Faster R-CNN [3]	Single	70.3	83	76.1	0.101 s
Proposed	Single	76.3	91.8	83.3	0.137 s
Proposed (competition mode)	Single	87.1	87.7	87.4	-

Figure 5. Result examples on ICDAR2013. Red rectangles are detection results by proposed method.

4.3. *Ablation Study*

We discuss the effectiveness of the proposed method by ablation study. There are four variations. The first is baseline Faster R-CNN. The second is Faster R-CNN with the proposed anchors (Anchor). The third is Faster R-CNN with SSP-RPNs (SSP-RPNs). The last is the proposed method with the proposed anchors and SSP-RPNs (Proposed). We used the same environment and hyperparameters for training all the variations.

We showed the results in Table 2. Compared to the baseline and Anchor, F-score improved by 6 points, which indicates the effectiveness of the proposed anchors. The anchor is an important factor to generate bounding-boxes in the RPN. Compared to the baseline and SSP-RPNs, F-score improved by 1.5 points. We confirmed that the scale sensitive module made detection effective. The proposed method was better than other methods. Therefore, both the proposed anchors and modules should be robust for text detection. The proposed method also improved the precision with a large margin. Thus, the proposed method learned to generate proposals by reducing negative proposals. Overall, the proposed method improved robustness with the help of the multiple RPNs.

Table 2. Results on ablation study.

Method	Anchor	SSP-RPNs	Recall	Precision	F-Score	Time
Baseline			70.28	82.99	76.11	0.101
Anchor	✓		77.21	88.35	82.40	0.103
SSP-RPNs		✓	70.26	86.63	77.86	0.125
Proposed	✓	✓	76.29	91.81	83.33	0.137

Subsequently, we discuss the detected bounding-boxes. Figure 6 shows that proposed method can utilize the receptive field and context. On the other hand, the base failed to enclose the texts entirely. The RPN in the baseline has 228 × 228 receptive fi which is smaller than the target text scale. We assumed that this failure was due less context. Compared to the baseline, the proposed method enclosed large-scale completely. The large receptive field of the proposed method extracted enough contex confirm the existence of large texts in image. Consequently, we achieved accurate detect

The third row in Figure 6 also shows the validity of Proposed. The anchor mo detected large texts, however, they are partial. This failure was caused by a small cont in target texts. On the other hand, the SSP-RPNs model and the proposed model detec large texts successfully. These results show that a horizontally long anchor is necessary Latin text detection. Besides, receptive field positively contributes to generating propos Thus, SSP-RPNs module is essential.

Figure 6. Detection examples in ablation study.

4.4. *Scale Sensitive Strategy*

In this section, we evaluate the SSP-RPNs module. Figure 7 showed that the outpu of each RPN, RoI-merge layer, and results. The upper row in Figure 7 is a tiny-scale te case. The RPN1 generated proposals fitted to the tiny text with high confidence, where RPN3 and RPN4 failed. After the RoI-merge layer, the proposals of RPN1 were selecte Consequently, detection succeeded in the final result. These results verified that RPN learned small texts. The lower row in Figure 7 is a large text case. The proposals of RPN were too small to enclose the entire text region. Whereas RPN3 and RPN4 generate proposals enclosing the whole text region. Consequently, the large texts were detected the final result.

Overall, each RPN learned to detect each suitable scale text corresponding to the receptive field sizes, i.e., RPN1 was optimized for small-scale, and RPN3 and 4 we optimized for large-scale. Therefore, these RPNs can help RPN2, which is in its origin position after conv5-3. Moreover, the RoI-merge layer is necessary for the SSP-RPN module to reject unnecessary proposals.

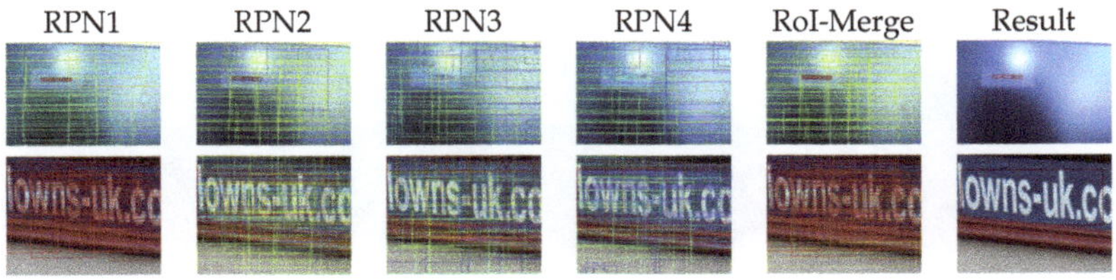

Figure 7. The outputs from each region proposal network (RPN). Red rectangles have high confidence, and blue rectangles have low confidence.

4.5. Failure Analysis

We analyzed the failure results of the proposed method. The failure examples are shown in Figures 8 and 9a–e show the proposals from each RPN and RoI-merge layer, (f) is the outputs of the classification by the Fast R-CNN and non-maximum suppression, (g) shows the final output, and (h) is some examples of the output feature map from conv5-3.

Figure 8 shows some text regions in the bottom-right image were not detected. The RPNs generated proposals of all the text regions, as well as RoI-merge layer. However, proposals were misclassified. Thus, some proposals were rejected by low confidence as the final output. As shown in Figure 8h, the bottom-right text regions were not activated well. To correct the proposed method, the classifier in the proposed method needs more training. The total loss is mostly occupied by the RPN losses, in Equation (1). Thus, we need to take a balance over L_{rpn_i} and $L_{fastrcnn}$.

(**a**) RPN1 (**b**) RPN2 (**c**) RPN3 (**d**) RPN4

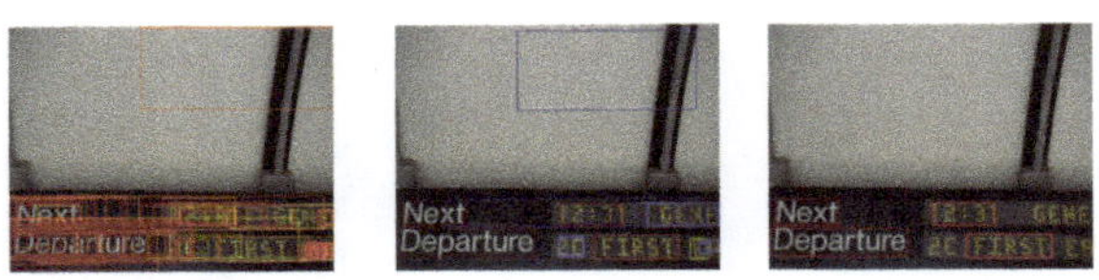

(**e**) RoI-merge layer (**f**) Output (**g**) Final Result

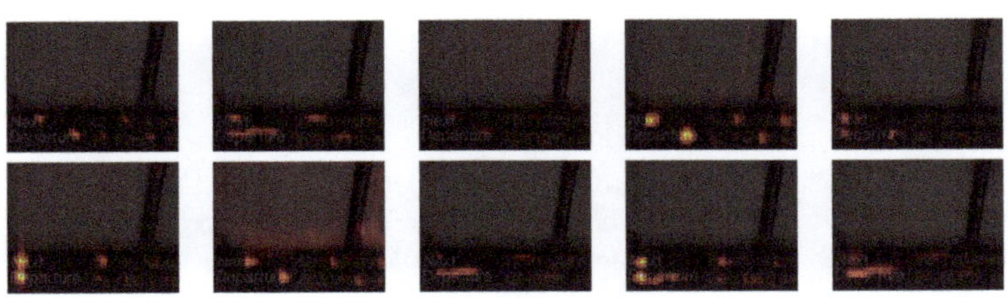

(**h**) conv5-3 activation map

Figure 8. Failure examples 1. (**f**) classification results of the proposals. Red and blue represent text and background, respectively. Activation maps in (**h**) are resized to the input size.

Moreover, we discuss on the case of Figure 9. The results contained the digit regions, however, they included large background regions. There are some proposals fitted to only digits. However, such proposals were misclassified. On the other hand, proposals with large background regions were classified to text with high confidence. According to (h),

background regions were activated. We can suppress the activation in the backgroun
assigning more weight to the classifier.

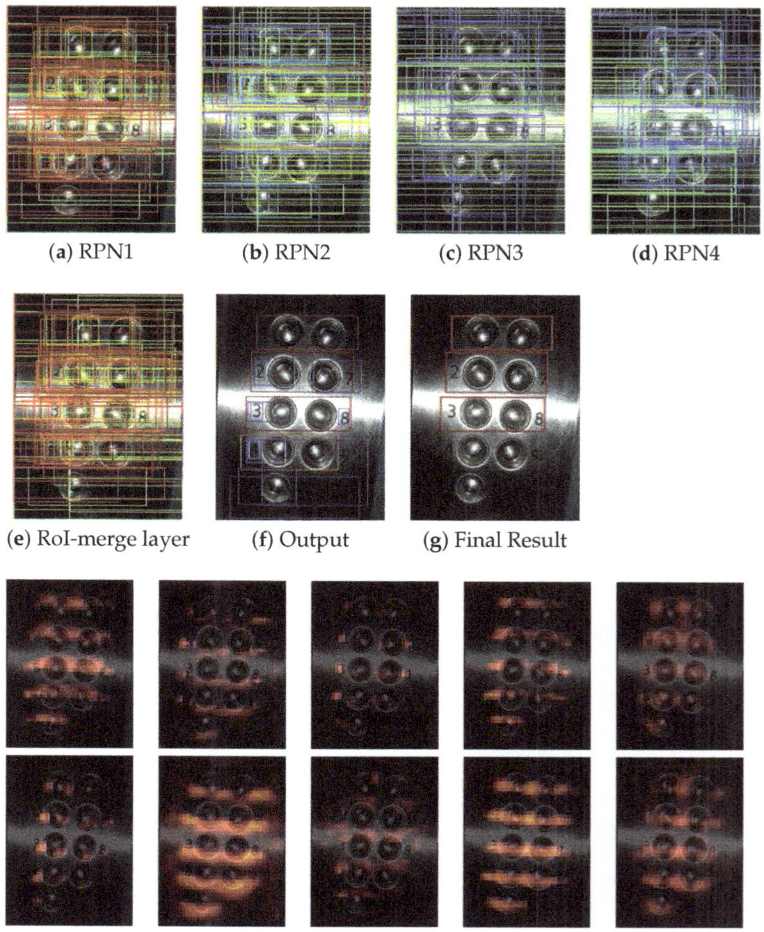

(**a**) RPN1 (**b**) RPN2 (**c**) RPN3 (**d**) RPN4

(**e**) RoI-merge layer (**f**) Output (**g**) Final Result

(**h**) conv5-3 activation map

Figure 9. Failure example 2.

5. Conclusions

We proposed a text detection method that is robust to text scales in natural scen images. The proposed method is based on the Faster R-CNN [3]. The main improv ment is to introduce multiple RPNs to detect texts from different resolution feature map We designed the anchors suitable for Latin text detection by the analysis on the thre datasets: COCO Text, Synth Text, and our dataset. We stress that these datasets are public and widely used in text detection studies. Thus, the proposed anchors ensure the genera ization capability. The experimental results show that the proposed method outperforme the Faster R-CNN at F-score with more than 7 points. Moreover, the proposed metho achieved comparable results to other methods. Therefore, we verified the effectiveness o the proposed method, especially for text scales.

Author Contributions: Conceptualization, Y.N. and T.M.; methodology, Y.N.; software, Y.N.; validation, Y.N. and T.M.; formal analysis, Y.N.; investigation, Y.N. and T.M.; resources, Y.N.; data curation, Y.N.; writing—original draft preparation, Y.N. and T.M.; writing—review and editing, Y.S. and S.O.; visualization, Y.N.; supervision, Y.S. and S.O.; project administration, S.O.; funding acquisition, S.O. All authors have read and agreed to the published version of the manuscript.

Funding: This work was partially supported by JSPS KAKENHI Grant Numbers 20H04201 and 18K19772.

Institutional Review Board Statement: Not applicable.

Informed Consent Statement: Not applicable.

Data Availability Statement: Publicly available datasets were analyzed in this study. This paper contains the links of the datasets.

Conflicts of Interest: The authors declare no conflict of interest.

erences

Huang, Z.; Lin, J.; Yang, H.; Wang, H.; Bai, T.; Liu, Q.; Pang, Y. An Algorithm Based on Text Position Correction and Encoder-Decoder Network for Text Recognition in the Scene Image of Visual Sensors. *Sensors* **2020**, *20*, 2942. [CrossRef]

Li, Z.; Zhou, Y.; Sheng, Q.; Chen, K.; Huang, J. A High-Robust Automatic Reading Algorithm of Pointer Meters Based on Text Detection. *Sensors* **2020**, *20*, 5946. [CrossRef] [PubMed]

Ren, S.; He, K.; Girshick, R.; Sun, J. Faster R-CNN: Towards Real-Time Object Detection with Region Proposal Networks. *IEEE Trans. Pattern Anal. Mach. Intell.* **2017**, *39*, 1137–1149. [CrossRef] [PubMed]

Redmon, J.; Divvala, S.; Girshick, R.; Farhadi, A. You Only Look Once: Unified, Real-Time Object Detection. In Proceedings of the IEEE Conference on Computer Vision and Pattern Recognition (CVPR), Las Vegas, NV, USA, 27–30 June 2016; pp. 779–788.

Redmon, J.; Farhadi, A. YOLO9000: Better, Faster, Stronger. In Proceedings of the IEEE Conference on Computer Vision and Pattern Recognition (CVPR), Honolulu, HI, USA, 21–26 July 2017; pp. 6517–6525.

Redmon, J.; Farhadi, A. YOLOv3: An Incremental Improvement. *arXiv* **2018**, arXiv:1804.02767.

Liu, W.; Anguelov, D.; Erhan, D.; Szegedy, C.; Reed, S.; Fu, C.Y.; Berg, A.C. SSD: Single Shot MultiBox Detector. In *Computer Vision—ECCV*; Leibe, B., Matas, J., Sebe, N., Welling, M., Eds.; Springer: Cham, Switzerland, 2016; pp. 21–37.

Lin, T.; Dollár, P.; Girshick, R.; He, K.; Hariharan, B.; Belongie, S. Feature Pyramid Networks for Object Detection. In Proceedings of the IEEE Conference on Computer Vision and Pattern Recognition (CVPR), Honolulu, HI, USA, 21–26 July 2017; pp. 936–944.

Kong, T.; Yao, A.; Chen, Y.; Sun, F. HyperNet: Towards Accurate Region Proposal Generation and Joint Object Detection. In Proceedings of the IEEE Conference on Computer Vision and Pattern Recognition (CVPR), Las Vegas, NV, USA, 27–30 June 2016; pp. 845–853.

He, K.; Zhang, X.; Ren, S.; Sun, J. Deep residual learning for image recognition. In Proceedings of the IEEE Conference on Computer Vision and Pattern Recognition (CVPR), Las Vegas, NV, USA, 27–30 June 2016; pp. 770–778.

Wang, H.; Wang, Q.; Gao, M.; Li, P.; Zuo, W. Multi-scale Location-aware Kernel Representation for Object Detection. In Proceedings of the IEEE Conference on Computer Vision and Pattern Recognition, Salt Lake City, UT, USA, 18–23 June 2018; pp. 1248–1257.

Nagaoka, Y.; Miyazaki, T.; Sugaya, Y.; Omachi, S. Text Detection by Faster R-CNN with Multiple Region Proposal Networks. In Proceedings of the 2017 14th IAPR International Conference on Document Analysis and Recognition (ICDAR), Kyoto, Japan, 9–15 November 2017; pp. 15–20.

Felzenszwalb, P.; McAllester, D.; Ramanan, D. A discriminatively trained, multiscale, deformable part model. In Proceedings of the Computer Vision and Pattern Recognition, Anchorage, AK, USA, 23–28 June 2008; pp. 1–8.

Ren, X.; Ramanan, D. Histograms of sparse codes for object detection. In Proceedings of the IEEE Conference on Computer Vision and Pattern Recognition, Portland, OR, USA, 23–28 June 2013; pp. 3246–3253.

Girshick, R.B.; Donahue, J.; Darrell, T.; Malik, J. Rich Feature Hierarchies for Accurate Object Detection and Semantic Segmentation. In Proceedings of the 2014 IEEE Conference on Computer Vision and Pattern Recognition, Columbus, OH, USA, 23–28 June 2014; pp. 580–587.

Uijlings, J.R.R.; van de Sande, K.E.A.; Gevers, T.; Smeulders, A.W.M. Selective Search for Object Recognition. *Int. J. Comput. Vis.* **2013**, *104*, 154–171. [CrossRef]

Girshick, R. Fast R-CNN. In Proceedings of the IEEE International Conference on Computer Vision (ICCV), Santiago, Chile, 7–13 December 2015.

Simonyan, K.; Zisserman, A. Very Deep Convolutional Networks for Large-Scale Image Recognition. In Proceedings of the International Conference on Learning Representations, San Diego, CA, USA, 7–9 May 2015.

Kong, T.; Sun, F.; Yao, A.; Liu, H.; Lu, M.; Chen, Y. Ron: Reverse connection with objectness prior networks for object detection. In Proceedings of the IEEE Conference on Computer Vision and Pattern Recognition, Honolulu, HI, USA, 21–26 July 2017; Volume 1, p. 2.

20. Ronneberger, O.; Fischer, P.; Brox, T. U-net: Convolutional networks for biomedical image segmentation. In Proceedings of th ternational Conference on Medical Image Computing and Computer-Assisted Intervention, Munich, Germany, 5–9 October 2 pp. 234–241.
21. Badrinarayanan, V.; Kendall, A.; Cipolla, R. Segnet: A deep convolutional encoder-decoder architecture for image segmenta *arXiv* **2015**, arXiv:1511.00561.
22. Yu, C.; Wang, J.; Peng, C.; Gao, C.; Yu, G.; Sang, N. Bisenet: Bilateral segmentation network for real-time semantic segmenta *arXiv* **2018**, arXiv:1808.00897.
23. Wang, K.; Babenko, B.; Belongie, S. End-to-end scene text recognition. In Proceedings of the International Conference Computer Vision, Barcelona, Spain, 6–13 November 2011; pp. 1457–1464.
24. Ozuysal, M.; Calonder, M.; Lepetit, V.; Fua, P. Fast Keypoint Recognition Using Random Ferns. *IEEE Trans. Pattern Anal. M Intell.* **2010**, *32*, 448–461. [CrossRef] [PubMed]
25. Felzenszwalb, P.F.; Huttenlocher, D.P. Pictorial Structures for Object Recognition. *Int. J. Comput. Vis.* **2005**, *61*, 55–79. [Cross
26. Wang, T.; Wu, D.J.; Coates, A.; Ng, A.Y. End-to-end text recognition with convolutional neural networks. In Proceedings of 21st International Conference on Pattern Recognition (ICPR), Tsukuba, Japan, 11–15 November 2012; pp. 3304–3308.
27. Milyaev, S.; Barinova, O.; Novikova, T.; Kohli, P.; Lempitsky, V. Image Binarization for End-to-End Text Understanding in Natu Images. In Proceedings of the 12th International Conference on Document Analysis and Recognition, Washington, DC, U 25–28 August 2013; pp. 128–132.
28. Opitz, M.; Diem, M.; Fiel, S.; Kleber, F.; Sablatnig, R. End-to-End Text Recognition Using Local Ternary Patterns, MSER and D Convolutional Nets. In Proceedings of the 11th IAPR International Workshop on Document Analysis Systems, Tours, Fra 7–10 April 2014; pp. 186–190.
29. Matas, J.; Chum, O.; Urban, M.; Pajdla, T. Robust wide-baseline stereo from maximally stable extremal regions. *Image Vis. Com* **2004**, *22*, 761–767. [CrossRef]
30. Jaderberg, M.; Simonyan, K.; Vedaldi, A.; Zisserman, A. Reading Text in the Wild with Convolutional Neural Networks. *In Comput. Vis.* **2016**, *116*, 1–20. [CrossRef]
31. Zitnick, C.L.; Dollár, P. Edge boxes: Locating object proposals from edges. In *European Conference on Computer Vision*; Spring Cham, Switzerland, 2014; pp. 391–405.
32. Dollár, P.; Appel, R.; Belongie, S.; Perona, P. Fast feature pyramids for object detection. *IEEE Trans. Pattern Anal. Mach. Int* **2014**, *36*, 1532–1545. [CrossRef] [PubMed]
33. Tian, S.; Pan, Y.; Huang, C.; Lu, S.; Yu, K.; Lim Tan, C. Text flow: A unified text detection system in natural scene imag In Proceedings of the IEEE International Conference on Computer Vision, Santiago, Chile, 7–13 December 2015; pp. 4651–46
34. Chen, X.; Yuille, A.L. Detecting and reading text in natural scenes. In Proceedings of the 2004 IEEE Computer Society Confere on Computer Vision and Pattern Recognition, Washington, DC, USA, 27 June–2 July 2004; Volume 2, p. II.
35. Liao, M.; Shi, B.; Bai, X.; Wang, X.; Liu, W. TextBoxes: A Fast Text Detector with a Single Deep Neural Network. *Proc. AAAI Co Artif. Intell.* **2017**, *31*, 4161–4167.
36. Tian, Z.; Huang, W.; He, T.; He, P.; Qiao, Y. Detecting text in natural image with connectionist text proposal network. In *Europe Conference on Computer Vision*; Springer: Cham, Switzerland, 2016; pp. 56–72.
37. Zhong, Z.; Jin, L.; Zhang, S.; Feng, Z. Deeptext: A unified framework for text proposal generation and text detection in natu images. *arXiv* **2016**, arXiv:1605.07314.
38. Szegedy, C.; Liu, W.; Jia, Y.; Sermanet, P.; Reed, S.; Anguelov, D.; Erhan, D.; Vanhoucke, V.; Rabinovich, A. Going deep with convolutions. In Proceedings of the IEEE Conference on Computer Vision and Pattern Recognition, Boston, MA, US 7–12 June 2015; pp. 1–9.
39. Tang, Y.; Wu, X. Scene text detection and segmentation based on cascaded convolution neural networks. *IEEE Trans. Ima Process.* **2017**, *26*, 1509–1520. [CrossRef] [PubMed]
40. Dai, Y.; Huang, Z.; Gao, Y.; Xu, Y.; Chen, K.; Guo, J.; Qiu, W. Fused text segmentation networks for multi-oriented scene te detection. In Proceedings of the 24th International Conference on Pattern Recognition (ICPR), Beijing, China, 20–24 August 20 pp. 3604–3609.
41. Lyu, P.; Yao, C.; Wu, W.; Yan, S.; Bai, X. Multi-oriented scene text detection via corner localization and region segmentatic In Proceedings of the IEEE Conference on Computer Vision and Pattern Recognition, Salt Lake City, UT, USA, 18–23 June 20 pp. 7553–7563.
42. Zhou, X.; Yao, C.; Wen, H.; Wang, Y.; Zhou, S.; He, W.; Liang, J. EAST: An efficient and accurate scene text detector. In Proceedin of the IEEE Conference on Computer Vision and Pattern Recognition (CVPR), Honolulu, HI, USA, 21–26 July 2017; pp. 2642–26
43. He, W.; Zhang, X.Y.; Yin, F.; Liu, C.L. Deep direct regression for multi-oriented scene text detection. *arXiv* **2017**, arXiv:1703.082
44. Shi, B.; Bai, X.; Yao, C. An end-to-end trainable neural network for image-based sequence recognition and its application to scer text recognition. *IEEE Trans. Pattern Anal. Mach. Intell.* **2017**, *39*, 2298–2304. [CrossRef] [PubMed]
45. Graves, A.; Fernández, S.; Gomez, F.; Schmidhuber, J. Connectionist temporal classification: Labelling unsegmented sequen data with recurrent neural networks. In *Proceedings of the 23rd International Conference on Machine Learning*; Association f Computing Machinery: New York, NY, USA, 2006; pp. 369–376.
46. Bušta, M.; Neumann, L.; Matas, J. Deep textspotter: An end-to-end trainable scene text localization and recognition framewor In Proceedings of the IEEE International Conference on Computer Vision (ICCV), Venice, Italy, 22–29 October 2017; pp. 2223–223

Li, H.; Wang, P.; Shen, C. Towards end-to-end text spotting with convolutional recurrent neural networks. In Proceedings of the IEEE Conference on Computer Vision and Pattern Recognition, Honolulu, HI, USA, 21–26 July 2017; pp. 5238–5246.
Liu, X.; Liang, D.; Yan, S.; Chen, D.; Qiao, Y.; Yan, J. FOTS: Fast Oriented Text Spotting with a Unified Network. In Proceedings of the IEEE Conference on Computer Vision and Pattern Recognition, Salt Lake City, UT, USA, 18–23 June 2018; pp. 5676–5685.
Veit, A.; Matera, T.; Neumann, L.; Matas, J.; Belongie, S. COCO-Text: Dataset and Benchmark for Text Detection and Recognition in Natural Images. *arXiv* **2016**, arXiv:1601.07140.
Gupta, A.; Vedaldi, A.; Zisserman, A. Synthetic Data for Text Localisation in Natural Images. In Proceedings of the IEEE Conference on Computer Vision and Pattern Recognition (CVPR), Las Vegas, NV, USA, 27–30 June 2016.
Karatzas, D.; Gomez-Bigorda, L.; Nicolaou, A.; Ghosh, S.; Bagdanov, A.; Iwamura, M.; Matas, J.; Neumann, L.; Chandrasekhar, V.R.; Lu, S.; et al. ICDAR 2015 competition on Robust Reading. In Proceedings of the 2015 13th International Conference on Document Analysis and Recognition (ICDAR), Tunis, Tunisia, 23–26 August 2015; pp. 1156–1160.
Nayef, N.; Yin, F.; Bizid, I.; Choi, H.; Feng, Y.; Karatzas, D.; Luo, Z.; Pal, U.; Rigaud, C.; Chazalon, J.; et al. ICDAR2017 Robust Reading Challenge on Multi-Lingual Scene Text Detection and Script Identification - RRC-MLT. In Proceedings of the 2017 14th IAPR International Conference on Document Analysis and Recognition (ICDAR), Kyoto, Japan, 9–15 November 2017; Volume 1, pp. 1454–1459.
Wolf, C.; Jolion, J.M. Object count/area graphs for the evaluation of object detection and segmentation algorithms. *Int. J. Doc. Anal. Recognit.* **2006**, *8*, 280–296. [CrossRef]

sensors

MDPI

Article

The Optimally Designed Variational Autoencoder Networks for Clustering and Recovery of Incomplete Multimedia Data

Xiulan Yu, Hongyu Li *, Zufan Zhang and Chenquan Gan

School of Communication and Information Engineering, Chongqing University of Posts and Telecommunications, Chongqing 400065, China; yuxl@cqupt.edu.cn (X.Y.); zhangzf@cqupt.edu.cn (Z.Z.); gancq@cqupt.edu.cn (C.G.)

* Correspondence: s160101070@stu.cqupt.edu.cn; Tel.: +86-187-2563-3001

Received: 29 January 2019; Accepted: 13 February 2019; Published: 16 February 2019

Abstract: Clustering analysis of massive data in wireless multimedia sensor networks (WMSN) has become a hot topic. However, most data clustering algorithms have difficulty in obtaining latent nonlinear correlations of data features, resulting in a low clustering accuracy. In addition, it is difficult to extract features from missing or corrupted data, so incomplete data are widely used in practical work. In this paper, the optimally designed variational autoencoder networks is proposed for extracting features of incomplete data and using high-order fuzzy c-means algorithm (HOFCM) to improve cluster performance of incomplete data. Specifically, the feature extraction model is improved by using variational autoencoder to learn the feature of incomplete data. To capture nonlinear correlations in different heterogeneous data patterns, tensor based fuzzy c-means algorithm is used to cluster low-dimensional features. The tensor distance is used as the distance measure to capture the unknown correlations of data as much as possible. Finally, in the case that the clustering results are obtained, the missing data can be restored by using the low-dimensional features. Experiments on real datasets show that the proposed algorithm not only can improve the clustering performance of incomplete data effectively, but also can fill in missing features and get better data reconstruction results.

Keywords: feature learning; incomplete multimedia data; fuzzy c-means; variational autoencoder

1. Introduction

The rapid development of communication technologies and sensor networks leads to the increase of heterogeneous data. The proliferation of these technologies in communication networks also has facilitated the development of the wireless multimedia sensor network (WMSN) [1]. Currently, multimedia data on WMSNs are successfully used in many applications, such as industrial control [2], target recognition [3] and intelligent traffic monitoring [4].

Nowadays, multimedia sensors produce a great deal of heterogeneous data, which require new models and technologies to process, particularly neural computing [5], to further promote the design and application of WMSNs [6,7]. However, heterogeneous networks and data are often very complex [8,9], which consist of structured data and unstructured data such as picture, voice, text, and video. Because heterogeneous data come from many input channels in the real world, these data are typical multimodal data, and there is a nonlinear relationship between them [10]. Different modes usually convey different information [11]. For example, images have many details, such as shadows, rich colors and complex scenes, and use titles to display invisible things like the names of objects in the image [12]. Moreover, different forms have complex relationships. In the real world, most multimedia data suffer from a lot of missing values due to sensor failures, measurement inaccuracy and network data transmission problems [13,14]. These features, especially incompleteness, lead to the widespread use of incomplete data in practical applications [15,16]. Lack of data values will affect the decision process of the application servers for specific tasks [17]. The resulting errors can be important for subsequent steps in data processing. Therefore, the recovery of data missing values is essential for processing big data in WMSNs.

As a fundamental technology of big data analysis, clustering divides objects into different clusters based on different similarity measures, making objects in the same cluster more similar to other objects in different groups [18,19]. They are commonly used to organize, analyze, communicate, and retrieve tasks [20]. Traditional data clustering algorithms focus on complete data processing, such as image clustering [21], audio clustering [22] and text clustering [23]. Recently, heterogeneous data clustering methods have been widely concerned by researchers [24–26]. In addition, many algorithms have been proposed—for example, Meng et al. optimized the unified objective function by an iterative process, and a spectral clustering algorithm is developed for clustering heterogeneous data based on graph theory [27]. Li et al. [28] proposed a high-order fuzzy c-means algorithm to extend the conventional fuzzy c-means algorithm from vector space to tensor space. A high-order possibilistic c-means algorithm based on tensor decompositions was proposed for data clustering in Internet of Things (IoT) systems [29]. These algorithms are effective to improve clustering performance for heterogeneous data. However, they can only obtain clustering results and lack further analysis of incomplete data low-dimensional features. Therefore, their performance is limited with the heterogeneous data in the WMSNs' big data environment. More importantly, other existing feature clustering algorithms do not consider data reconstruction and missing data. WMSN systems require different modern data analysis methods, and deep learning (DL) has been actively applied in many applications due to its strong data feature extraction ability [30]. Deep embedded clustering (DEC) learns to map from data space to low-dimensional feature space, where it optimizes the clustering objectives [31]. Ref. [32] shows the feature representation ability of variational autoencoder (VAE). VAE learns the multi-faceted structure of data and achieves high clustering performance [33]. In addition, VAE has a strong ability in feature extraction and reconstruction, and it can be a good tool for handling incomplete data.

Aiming at this research object, the variational autoencoder based high-order fuzzy c-means (VAE-HOFCM) algorithm is presented to cluster and reconstruction incomplete data in WMSNs in this paper. It can effectively cluster complete data and incomplete data and get better reconstruction results. VAE-HOFCM is mainly composed of three steps: feature learning and extraction, high-order clustering, and data reconstruction. First, the feature learning network is improved by using a variational autoencoder to learn the feature of incomplete data. To capture nonlinear correlations of different heterogeneous data, tensors are applied to form a feature representation of heterogeneous data. Then, the tensor distance is used as the distance measure to capture the unknown distribution of data as much as possible in the clustering process. The results of feature clustering and VAE output both affect the final clustering results. Finally, in the case of clustering results, the missing data can be restored by the low-dimensional features.

The rest of the paper is organized as follows: Section 2 presents related work to this paper. The proposed algorithm is illustrated in Section 3, and experimental results and analysis are described in Section 4. Finally, the whole paper is concluded in the last section.

2. Preliminaries

This section describes the variational autoencoder (VAE) and the fuzzy c-means (FCM), which will be useful in the sequel.

2.1. Variational Autoencoder

The variational autoencoder, which is a new method for nonlinear dimensionality reduction, is a great case of combining probability plots with deep learning [34,35]. Consider a dataset $X = \{x_1, x_2, ..., x_N\}$ which consists of N independent and identically distributed samples of continuous or discrete variables x. To generate target data x from hidden variable z, two blocks are used: encoder block and decoder block. Suppose that z is generated by some prior normal distribution $p_\theta = N\left(\mu, \sigma^2\right)$.

The true posterior density $p_\theta\left(z\left|x\right.\right)$ is intractable. Approximate recognition model $q_\phi\left(z\left|x\right.\right)$ as a probabilistic encoder. Similarly, refer to $p_\theta\left(x\left|z\right.\right)$ as a probability decoder because, given the code z, it produces a distribution over the possible corresponding value x. The parameters θ and ϕ are used to represent the structure and weight of the neural network used. These parameters are adjusted as part of the VAE training process and are considered constant later. Minimize the true posterior approximation of the KL divergence (Kullback–Leibler Divergence). When the divergence of KL is zero, $p_\theta\left(z\left|x\right.\right) = q_\phi\left(z\left|x\right.\right)$. Then, the true posterior distribution can be obtained. The KL divergence of approximation from the true posterior $D_{KL}\left(q_\phi\left(z\left|x\right.\right)\|p_\theta\left(z\left|x\right.\right)\right)$ can be formulated as:

$$\begin{aligned}\left(q_\phi\left(z\left|x\right.\right)\|p_\theta\left(z\left|x\right.\right)\right) &= \int_{-\infty}^{\infty} q_\phi\left(z\left|x\right.\right)\log\frac{q_\phi(z|x)}{p_\theta(z|x)}dz \\ &= \log p_\theta\left(x\right) + D_{KL}\left(q_\phi\left(z\left|x\right.\right)\|p_\theta\left(z\right)\right) - E_{q_\phi(z|x)}\left[\log p_\theta\left(x\left|z\right.\right)\right] \\ &\geq 0,\end{aligned} \tag{1}$$

which can also be written as:

$$\log p_\theta\left(x\right) \geq -D_{KL}\left(q_\phi\left(z\left|x\right.\right)\|p_\theta\left(z\right)\right) + E_{q_\phi(z|x)}\left[\log p_\theta\left(x\left|z\right.\right)\right]. \tag{2}$$

The right half of the inequality is called the variational lower bound on the marginal likelihood of data x, and can be written as:

$$L\left(\theta,\phi;x\right) \geq -D_{KL}\left(q_\phi\left(z\left|x\right.\right)\|p_\theta\left(z\right)\right) + E_{q_\phi(z|x)}\left[\log p_\theta\left(x\left|z\right.\right)\right]. \tag{3}$$

The second term $E_{q_\phi(z|x)}\left[\log p_\theta\left(x\left|z\right.\right)\right]$ requires estimation by sampling. A differentiable transformation $g_\phi\left(x,\varepsilon\right)$ of an auxiliary noise variable ε is used to reparameterize the approximation $q_\phi\left(z\left|x\right.\right)$. Then, form a Monte Carlo estimates of $E_{q_\phi(z|x)}\left[\log p_\theta\left(x\left|z\right.\right)\right]$:

$$E_{q_\phi(z|x)}\left[\log p_\theta\left(x\left|z\right.\right)\right] = \frac{1}{M}\sum_{m=1}^{M}\log p_\theta\left(x\left|z^m\right.\right), \tag{4}$$

where $z^m = g_\phi\left(x,\varepsilon^m\right) = \mu + \varepsilon^m \odot \sigma$, $\varepsilon^m \sim N\left(0,I\right)$ and m denotes the number of samples.

2.2. Fuzzy C-Means Algorithm (FCM)

The fuzzy c-means algorithm (FCM) is a typical soft clustering technique [36,37]. Given a dataset $X = \{x_1, x_2, \ldots, x_N\}$ with N objects and m observations, fuzzy partition of set X into predefined cluster number c and the number of clustering centers denoted by $V = \{v_1, v_2, \ldots, v_c\}$. Their membership functions are defined as $u_{ik} = u_{v_i}\left(x_k\right)$, in which u_{ik} denotes the membership of x_k towards the i th clustering center and c denotes. FCM is defined by a $c \times m$ membership matrix $U = \{u_{ik}\left|1 \leq i \leq c; 1 \leq k \leq m\right.\}$. FCM minimizes the following objective function [38,39] to calculate the membership matrix U and the clustering centers V:

$$J_m\left(U,V\right) = \sum_{k=1}^{n}\sum_{i=1}^{c}\left(u_{ik}\right)d^2\left(x_k, v_i\right), \tag{5}$$

where every u_{ik} belongs to the interval (0,1), the summary of all the u_{ik} belonging to the same point is one ($\sum_{i=1}^{c} u_{ik} = 1$). In addition, none of the fuzzy clusters is empty, neither do any contain all the data

$0 < \sum_{k=1}^{m} u_{ik} < m, 1 \leq i \leq c$. Update the membership matrix and clustering centers by minimizing Equation (5) via the Lagrange multipliers method:

$$u_{ik} = \frac{1}{\sum\limits_{j=1}^{c} \left(d_{ik}/d_{jk}\right)^{1/(m-1)}}, \tag{6}$$

$$v_i = \frac{\sum_{k=1}^{n} u_{ik}^{m} x_k}{\sum_{k=1}^{n} u_{ik}^{m}}. \tag{7}$$

In the traditional FCM algorithm, d_{ik} denotes the Euclidean distance between x_i and v_k, and d_{jk} denotes the Euclidean distance between x_j and v_k.

3. Problem Formulation and Proposed Method

Consider a dataset $X = \{x_1, x_2, \ldots x_N\}$ with N objects. Each object is represented by m observations, in the form of $Y = \{y_1, y_2, \ldots, y_m\}$. The purpose of data clustering is to divide datasets into several similar classes based on similarity measure, so that objects in the same cluster have great similarity and are easy to be analyzed. Multimedia data cluster tasks bring many problems and challenges, especially for missing or damaged data. Key challenges are discussed in three areas as below.

1. Learning the features of incomplete data: feature extraction and analysis are the basic steps of clustering. In general, many feature extraction methods, such as machine learning and deep learning, have been successfully applied to image, text, and audio feature learning. However, the current algorithm focuses on feature learning and extraction of high quality data. In other words, they can not effectively extract the features of lossy data. Therefore, feature learning of incomplete data is the primary problem of heterogeneous data clustering.
2. Clustering in feature space: an important feature of large-scale multimedia data is its diversity, which means that large-scale data sources are diverse, including structured, unstructured data and semi-structured data from a large number of sources. In particular, a large number of objects in large data sets are multi-model. For example, web pages usually contain both images and text. Each mode of multimodal object has its own characteristics, which leads to the complexity of data. Therefore, the feature representation of multimedia data is significant in cluster tasks.
3. Filling missing values to reconstruct data: in wireless multimedia sensor networks, reliable data transmission is critical to provide the ideal quality of network-based services. However, multimedia data transmission may not be successful due to different reasons such as sensory errors, connection errors, or external attacks. These problems can result in incomplete data and degrade the performance of WMSNS applications. After feature extraction and cluster analysis, it is very important to recover missing data from the sensor network.

3.1. Description of the Proposed Method

The variational autoencoder based high-order fuzzy c-means (VAE-HOFCM) algorithm is divided into three stages: unsupervised feature learning, high-order feature clustering, and data reconstruction. Architecture of the proposed method is shown in Figure 1.

To learn the features of incomplete multimedia data, the original data set is divided into two different subsets X_c and X_{inc}. Samples in subset X_c have no missing values while each sample contains some missing values in subset X_{inc}.

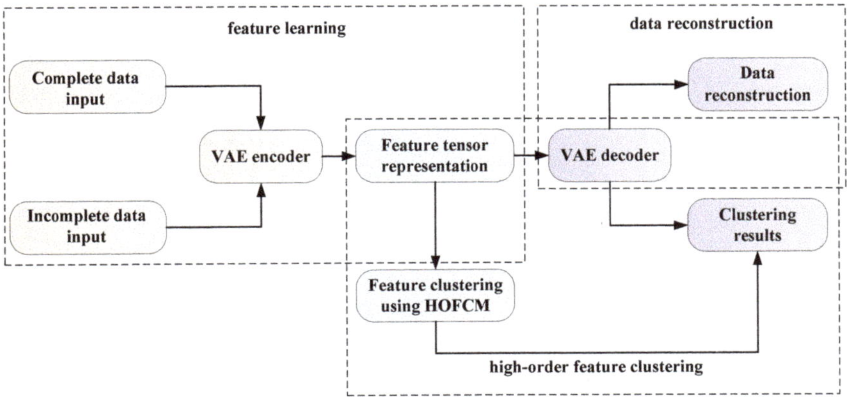

Figure 1. Architecture of the proposed method.

3.2. Feature Learning Network Architecture

For trained variational autoencoder, $q_\phi(z|x)$ will be very close to $p_\theta(z|x)$, so the encode network can reduce the dimensionality of the real dataset $X = \{x_1, x_2, ..., x_N\}$ and obtain low-dimensional distribution. In this case, the potential variables may get better results than the traditional dimensionality reduction methods. When the improved VAE model is obtained, the encode network is used to learn the potential feature vectors of missing value sample $z = Encoder(x) \sim q_\phi(z|x)$. The decode network is then used to decode the vector z to generate the original sample $\bar{x} = Decoder(z) \sim p_\theta(x|z)$.

According to the original VAE and to build a better generation model, convolution kernels are added to the encoder. There is a variational constraint on the latent variable z, that is, z obeys the Gauss distribution. Here, each x_i $(1 \leq i \leq N)$ is fitted with an exclusive normal distribution. Sample z is then extracted from the exclusive distribution, since z_i is sampled from the exclusive x_i distribution, the original sample x_i can be generated through a decoder network. The improved VAE model is shown in Figure 2.

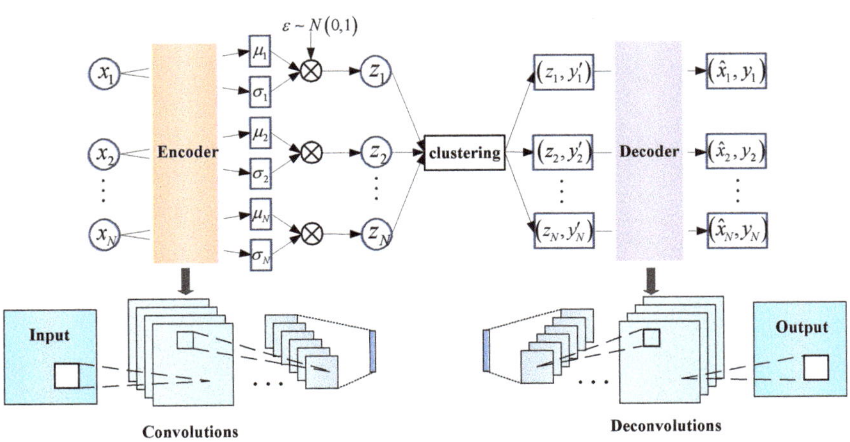

Figure 2. The improved VAE model.

In general, assume that $q_\phi(z)$ is the standard normal distribution, $q_\phi(z|x)$, $p_\theta(x|z)$ are the conditional normal distribution, and then plug in the calculation to get the normal loss of VAE, where z is a continuous variable representing the coding vector, and y is a discrete variable that represents a category. If z is directly replaced in the formula with (z, y), the loss of the clustered VAE is obtained:

$$D_{KL}\left(q_\phi(z,y|x)\,\|p_\theta(z,y|x)\right) = \int_{-\infty}^{\infty} q_\phi(z,y|x)\log\frac{q_\phi(z,y|x)}{p_\theta(z,y|x)}dz. \tag{8}$$

Set the scheme as: $q_\phi(z,y|x) = q_\phi(y|z)\,q_\phi(z|x)$, $p_\theta(x|z,y) = p_\theta(x|z)$, $p_\theta(z,y) = p_\theta(z|y)\,p_\theta(y)$. Substituting them into Equation (8) and it can be simplified as follows:

$$E_{q_\phi(x)}\left[-\log p_\theta(x|z) + \sum_y q_\phi(y|z)D_{KL}\left(q_\phi(z|x)\,\|p_\theta(z|y)\right) + D_{KL}\left(q_\phi(y|z)\,\|p_\theta(y)\right)\right], \tag{9}$$

where the first term $-\log p_\theta(x|z)$ wants the reconstruction error to be as small as possible, that is, z keeps as much information as possible. $\sum_y q_\phi(y|z)D_{KL}\left(q_\phi(z|x)\,\|p_\theta(z|y)\right)$ plays the role of clustering. In addition, $D_{KL}\left(q_\phi(y|z)\,\|p_\theta(y)\right)$ makes the distribution of each class as balanced as possible; there will not be two nearly overlapping situations. The above equation describes the coding and generation process:

- Sampling to x from the original data, coding feature z can then be obtained by $q_\phi(z|x)$. Then, the coding feature is classified by classifier $q_\phi(y|z)$ to obtain the classification.
- Select a category y from distribution $p_\theta(y)$, select a random hidden variable z from distribution $p_\theta(z|y)$, and then decode the original sample through generator $p_\theta(x|z)$.

The VAE is outlined in Algorithm 1.

Algorithm 1 Variational Autoencoder Optimization.

Input: Training set $X = \{x_t\}_{t=1}^N$, corresponding labels $Y = \{y_t\}_{t=1}^N$, loss weight $\lambda_1, \lambda_2, \lambda_3$.

Output: VAE parameters θ,ϕ.

1: **Initialization:** random initialized θ_0, ϕ_0.
2: **Repeat:** Sample x_t in the minibatch.
3: $\mu_{z_t} = Encoder(x_t) \sim q_\phi(z|x)$
4: Sample: $z_t \leftarrow \mu_{z_t} + \varepsilon \odot \sigma_z, \varepsilon \sim \mathrm{N}(0, \mathrm{I})$
5: $\mu_{x_t} = Decoder(z_t) \sim p_\theta(x|z)$
6: Compute reconstruction loss: $L_{rec} = -\log p_\theta(x_t|z_t)$.
7: Compute regularization loss: $L_{reg} = D_{KL}\left(q_\phi(y_t|z_t)\,\|p_\theta(y_t)\right)$.
8: Compute clustering loss: $L_{cls} = \sum_y q_\phi(y_t|z_t)D_{KL}\left(q_\phi(z_t|x_t)\,\|p_\theta(z_t|y_t)\right)$.
9: Fuse the three loss: $L(\theta,\phi) = \lambda_1 L_{rec}(\theta,\phi) + \lambda_2 L_{reg}(\theta,\phi) + \lambda_3 L_{cls}(\theta,\phi)$.
10: Back-propagate the gradients.
11: **Until** maximum iteration reached.

3.3. Variational Autoencoder Based High-Order Fuzzy C-Means Algorithm

Variational autoencoder gets the low-dimensional features and initial clustering results of data by feature learning. Then, the final clustering results will be optimized by the FCM algorithm clustering results. Traditional FCM work in vector space. It is better to use higher-order tensor to represent the feature of data because the tensor distance can capture the correlation in the high-order tensor space and measures the similarity between two higher-order complex data samples. Given an N-order tensor $X \in R^{I_1 \times I_2 \times \ldots \times I_N}$, x is denoted as the vector form representation of X, and the element $X_{i_1 i_2 \ldots i_N (1 \le i_j \le I_j, 1 \le j \le N)}$ in X is corresponding to x_l. That is, the N element in X is $l = i_1 + \sum_{j=2}^{N} \prod_{t=1}^{j-1} I_t$. Then, the tensor distance between two N-order tensors is defined as:

$$d_{td} = \sqrt{\sum_{l,m=1}^{I_1 \times I_2 \times \ldots \times I_N} g_{lm} (x_l - y_l)(x_m - y_m)} = \sqrt{(x-y)^T G (x-y)}, \quad (10)$$

where g_{lm} is the metric coefficient and used to capture the correlations between different coordinates in the tensor space, which can be calculated by:

$$g_{lm} = \frac{1}{2\pi\delta^2} \exp\left\{-\frac{\|p_l - p_m\|_2^2}{2\delta^2}\right\}, \quad (11)$$

where $\|p_l - p_m\|_2$ is defined as:

$$\|p_l - p_m\|_2 = \sqrt{(i_1 - i_1')^2 + \cdots + (i_N - i_N')^2}. \quad (12)$$

Minimizing the objective function of high-order fuzzy c-means algorithm:

$$J_m(U,V) = \sum_{k=1}^{n} \sum_{i=1}^{c} (u_{ik}) d_{td}^2. \quad (13)$$

To update the membership value u_{ik}, we differentiate with respect to u_{ik}, as follows:

$$\begin{aligned} \frac{\partial J_m(U,V)}{\partial u_{ij}} &= \frac{\partial\left((u_{ik})^m d_{td}^2(x_k, v_i)\right)}{\partial u_{ij}} \\ &= m \cdot (u_{ij})^{m-1} d_{td}^2(x_j, v_i). \end{aligned} \quad (14)$$

Setting Equation (14) to 0, u_{ik} is calculated:

$$u_{ik} = \frac{1}{\sum_{j=1}^{c} \left(\frac{d_{(td)ik}}{d_{(td)jk}}\right)^{1/(m-1)}}. \quad (15)$$

Then, the equation for updating v_i is obtained:

$$v_i = \frac{\sum_{j=1}^{n} u_{ij}^m x_j}{\sum_{j=1}^{n} u_{ij}^m}. \quad (16)$$

For each iteration, this operation requires $O(c \times n)$, so the total computational complexity of k iterations is $O(kc \times n)$. From the above, the VAE-HOFCM algorithm can be described as Algorithm 2:

Algorithm 2 The VAE-HOFCM algorithm.

Input: $X = \{x_1, x_2, ..., x_n\}$

Output: $U = (u_{ij})$ and $V = (v_i)$.

Initialize $X = \{x_1, x_2, ..., x_n\}$ randomly.
Perform Algorithm 1 to calculate low dimensional representation of dataset X: $x = Encoder\,(x_n)$
for $iteration = 1, 2, \ldots, \max iter$
  for: $\mathrm{i} = 1, 2, \ldots, c$
  $v_i = \frac{\sum_{j=1}^{n} u_{ij}^{m} x_j}{\sum_{j=1}^{n} u_{ij}^{m}}$
  for: $\mathrm{i} = 1, 2, \ldots, c$
    for: $\mathrm{j} = 1, 2, \ldots, c$
    $u_{ij} = \frac{1}{\sum\limits_{j=1}^{c} \left(\frac{d_{(td)ik}}{d_{(td)jk}} \right)^{1/(m-1)}}$
$(x, y) = Decoder\,(z_t)$.
Obtain the modified clustering results using the u_{ij}.

By comparing the steps of the HOFCM algorithm, VAE-HOFCM can restore incomplete data simultaneously in the clustering process. Equally, the VAE-HOFCM algorithm has a total time complexity of $O\,(kc \times n)$. However, before that, it needs to train the variational autoencoder network.

4. Experiments

This section evaluates the performance of the proposed VAE-HOFCM algorithm on three representative datasets. To show the effectiveness of VAE-HOFCM, the unsupervised clustering accuracy (ACC) and adjusted rand index (ARI) for verification are adopted. ACC is calculated by:

$$ACC = \max_{m} \frac{\sum_{i=1}^{n} 1\,\{l_i = m\,(c_i)\}}{n}, \tag{17}$$

where l_i and c_i indicate the ground-truth label and the cluster assignment produced by the algorithm, respectively. m ranges overall possible one-to-one mappings between clusters and labels. ARI is used to measure the agreement between two possibilistic partitions of a set of objects, where U denotes the true labels of the objects in datasets, and U' denotes a cluster generated by a specific algorithm. A higher value of $ARI\,(U, U')$ represents that the algorithm has more accurate clustering results.

To study the performance and generality of different algorithms, experiments are performed on three datasets:

- MNIST: The MNIST dataset consists of 70,000 hand-written digits of 28-by-28 pixel size. The digits are centered and and the size is standardized.
- STL-10: A dataset consists of 96-by-96 color images. It contains 13,000 labeled images and 100,000 unlabeled images.
- NUS-WIDE: The NUS-WIDE dataset consists of 269,648 images and can be downloaded from Flickr.com, a famous photo-sharing website.

4.1. Experimental Results on Complete Datasets

This section evaluates the performance of variational autoencoder based high-order fuzzy c-means algorithm (VAE-HOFCM) in clustering compared to other algorithms. The input dimensions of these three datasets are 784, 3072 and 500, respectively. The dimension of VAE hidden layer is set as 25, and the number of training iterations of the training set as 50. After obtaining the low-dimensional features, start clustering, and the membership factor is set as 2.5. Then, the required clustering center is calculated and the final normalized membership matrix U is returned to obtain the clustering result.

The clustering results are shown in Tables 1 and 2. Table 1 displays the optimal performance of unsupervised clustering accuracy of each algorithm. For MNIST data clustering class, the proposed VAE-HOFCM algorithm has achieved the highest accuracy of 85.54%. Compared with VAE clustering, the VAE-HOFCM encoder training time and cluster running time sum is slightly more than the former, but the clustering accuracy is improved. Then, the clustering performance and running time of VAE-HOFCM algorithm are generally better than traditional clustering algorithms, such as k-means and fuzzy c-means. Since the dimension of STL-10 dataset is higher and the information content is larger, the operation time of extracting features and clustering is relatively long. However, the proposed algorithm still gets the best running results. Visual features and text features are extracted from the NUS-WIDE dataset, and then these features are connected to form feature vectors. Finally, the feature vectors are clustered. The clustering results show the performance of the proposed algorithm.

Table 1. Clustering accuracy of ACC.

Algorithm/Dataset	MNIST	STL-10	NUS-WIDE
k-means	53.49%	28.40%	81.51%
HOPCM	80.34%	33.12%	92.75%
VAE	84.20%	35.48%	93.32%
DEC	84.31%	35.90%	93.75%
VAE-HOFCM	85.54%	36.44%	95.14%

Table 2 shows the clustering results in terms of $ARI\ (U, U')$, VAE-HOFCM produces high value than other algorithms in most cases. K-means usually has the worst performance and the longest running time, whereas VAE and DEC achieve the better result than HOPCM. ARI is not used as an indicator in the STL-10 dataset because the value may be negative in the case of clustering accuracy.

Table 2. Clustering accuracy of ARI.

Algorithm/Dataset	MNIST	STL-10	NUS-WIDE
k-means	0.41	-	0.74
HOPCM	0.69	-	0.89
VAE	0.75	-	0.90
DEC	0.76	-	0.90
VAE-HOFCM	0.78	-	0.92

There are two reasons for the results of these results in terms of ACC and ARI. On the one hand, HOFCM integrates the learning characteristics of different modes, uses the cross product to model the nonlinear correlation under various modes, and uses the tensor distance as a measure to capture the high-dimensional distribution of multimedia data. On the other hand, VAE successfully learns low-dimensional features and achieves the best performance in feature dimension reduction and clustering accuracy.

VAE has good data clustering and data generation performance. Feature extraction is carried out by the VAE to reduce the dimension to two dimensions. These categories have clear boundaries as shown in Figure 3, indicating that the VAE has effectively extracted low-dimensional features. This proves that the VAE has strong data feature expression ability.

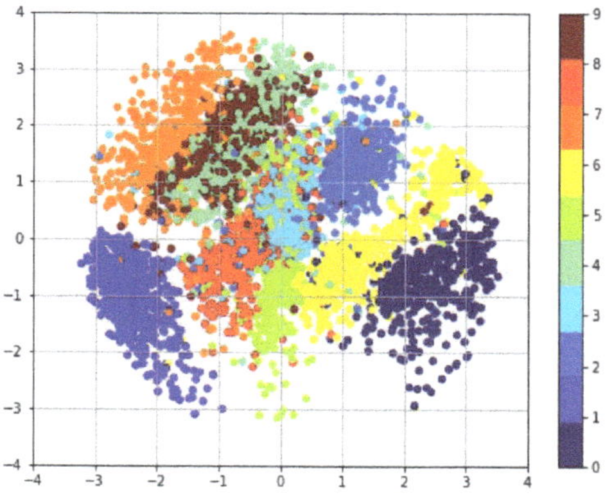

Figure 3. Visual analysis of MNIST datasets.

To obtain better performance in the three constraints of data feature dimension, clustering performance and reconstruction quality, the quality of data reconstruction in different dimensions is compared. Figure 4 shows the reproduction performance of learning generation models for different dimensions. When the latent space is set at 25, this method can obtain a good reconstruction quality.

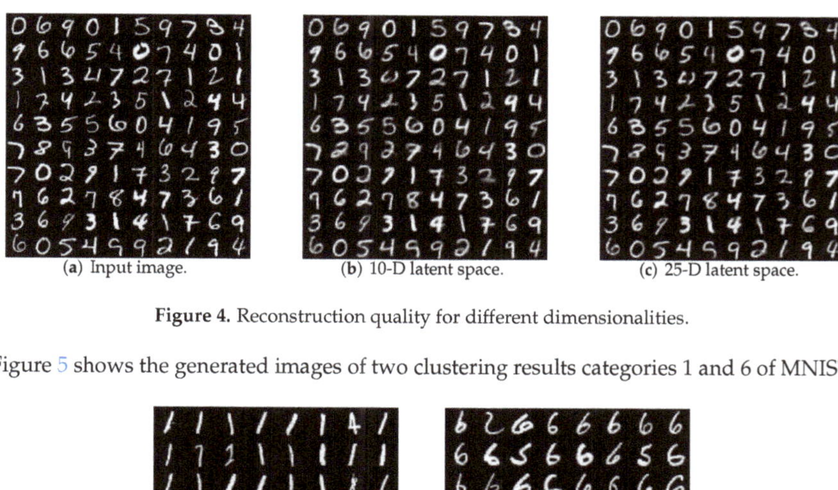

(a) Input image. (b) 10-D latent space. (c) 25-D latent space.

Figure 4. Reconstruction quality for different dimensionalities.

Figure 5 shows the generated images of two clustering results categories 1 and 6 of MNIST.

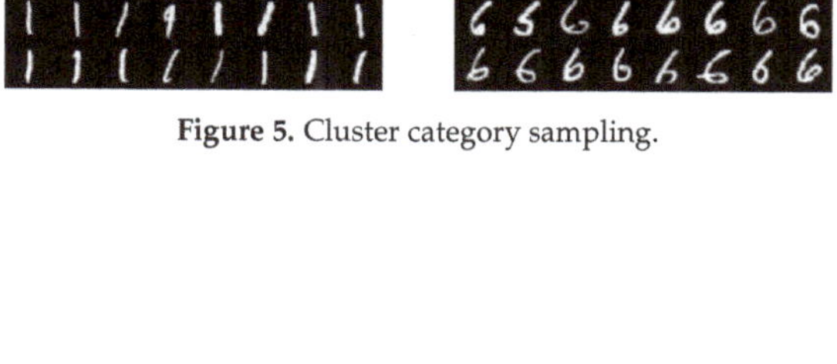

Figure 5. Cluster category sampling.

4.2. Experimental Results on Incomplete Data Sets

To estimate the robustness of the proposed algorithm, each dataset is divided into complete datasets and incomplete datasets. Now, incomplete datasets are used for simulation analysis. Since clustering performance depends on the number of missing values, six miss rates are set, which are 5%, 10%, 15%, 20%, 25% and 30%, respectively.

Figure 6 shows the clustering results accuracy of ACC with the increase of the missing ratio on the MNIST dataset and NUS-WIDE dataset. Figure 7 shows the average values of ARI with the increase of the missing ratio on the MNIST dataset and NUS-WIDE dataset. The results show that the increase of missing rate will lead to the decrease of clustering accuracy. However, the proposed algorithm still has a high accuracy because VAE successfully extracts incomplete data features and reduces the difference with the incomplete data features.

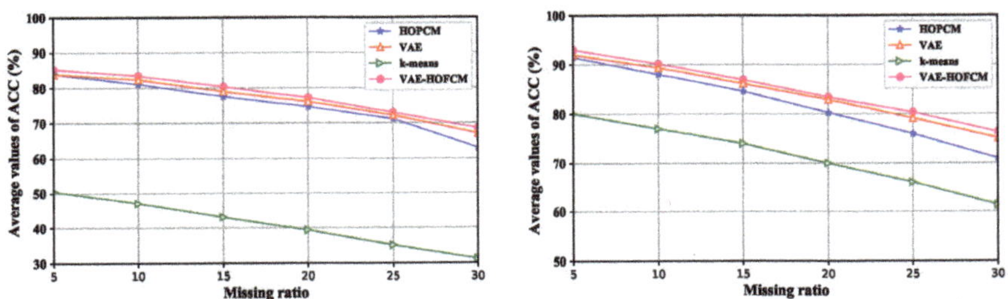

(**a**) Clustering accuracy of ACC on the MNIST dataset. (**b**) Clustering accuracy of ACC on the NUS-WIDE dataset.

Figure 6. Clustering accuracy of ACC.

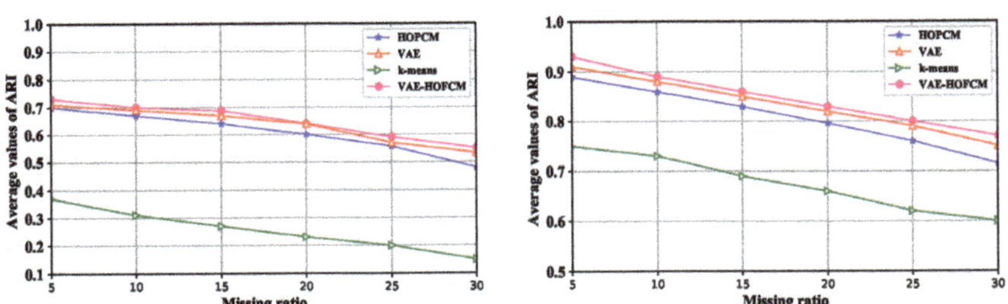

(**a**) Clustering accuracy of ARI on the MNIST dataset. (**b**) Clustering accuracy of ARI on the NUS-WIDE dataset.

Figure 7. Clustering accuracy of ARI.

According to Figures 6 and 7, with the increase of missing rate, the average value of ACC and ARI would decrease, which indicates that the missing rate destroys the original data content, leading to the decrease of clustering accuracy. The average ACC and ARI values based on the VAE-HOFCM algorithm are significantly higher than those of the other three methods at the six missing rates. Therefore, VAE-HOFCM clustering has the best performance, indicating that VAE-HOFCM is also effective for clustering incomplete data.

Then, data with different missing rates are reconstructed, as shown in Figure 8. Inputs are incomplete data with different missing rates, and the output are recovered data using VAE. The reconstruction results show that the proposed algorithm not only improves the clustering accuracy, but also ensures that the data can be reconstructed with high quality.

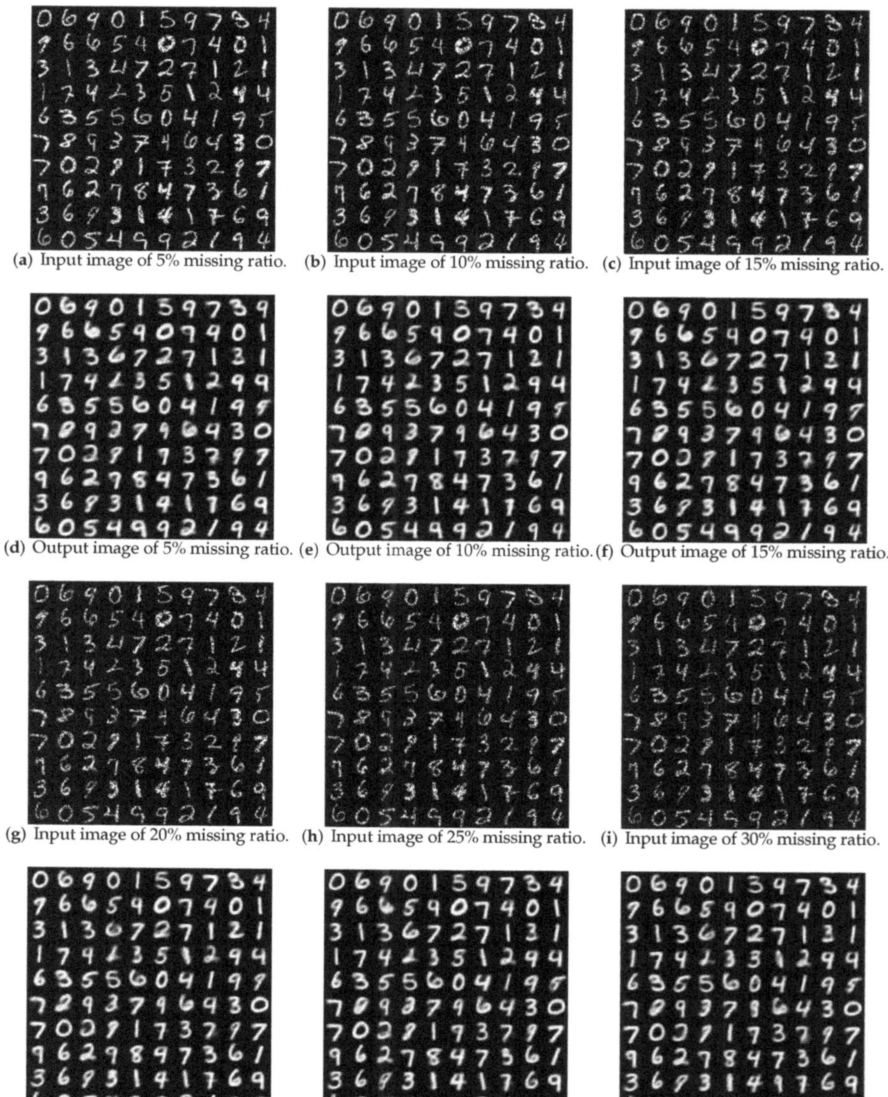

(**a**) Input image of 5% missing ratio. (**b**) Input image of 10% missing ratio. (**c**) Input image of 15% missing ratio.

(**d**) Output image of 5% missing ratio. (**e**) Output image of 10% missing ratio. (**f**) Output image of 15% missing ratio.

(**g**) Input image of 20% missing ratio. (**h**) Input image of 25% missing ratio. (**i**) Input image of 30% missing ratio.

(**j**) Output image of 20% missing ratio. (**k**) Output image of 25% missing ratio. (**l**) Output image of 30% missing ratio.

Figure 8. Reconstruction quality for different dimensionalities.

The variational auto-coder also has the function of de-noising. As shown in Figure 9, noise is added into the input data to enable VAE to effectively de-noise and restore the original input image.

(a) Original input image. (b) Input image with noise. (c) Restored image via VAE.

Figure 9. Reconstruction quality for noise data.

5. Conclusions

In this paper, a VAE-HOFCM algorithm, which can improve the performance of multimedia data clustering, has been proposed. Unlike many existing technologies, the VAE-HOFCM algorithm learns the data features by designing an improved VAE network, and uses a tensor based FCM algorithm to cluster the data features in the feature space. In addition, VAE-HOFCM captures as many features of high quality multimedia data and incomplete multimedia data as possible. In experiments, the performance of the proposed scheme has been evaluated on three heterogeneous datasets, MNIST, STL-10 and NUS-WIDE. Compared with traditional clustering algorithms, the results show that VAE can achieve a high compression rate of data samples, save memory space significantly without reducing clustering accuracy, and enable low-end devices in wireless multimedia sensor networks to achieve clustering of large data. In addition, VAE can effectively fill the missing data and generate the specified data at the terminal, so that the incomplete data can be better utilized and analyzed. Although VAE needs to be trained well, the sum time of training and clustering is still less than most clustering algorithms. Therefore, when performing clustering tasks on low-end equipment with limited computing power and memory space, trained VAE-HOFCM can be adopted.

Author Contributions: Conceptualization, X.Y. and Z.Z.; Data curation, C.G.; Formal analysis, X.Y. and H.L.; Funding acquisition, X.Y.; Investigation, H.L.; Supervision, Z.Z.; Validation, H.L. and C.G.; Visualization, Z.Z. and C.G.; Writing-original draft, X.Y. and H.L.; Writing-review and editing, X.Y., Z.Z. and C.G.

Funding: This work is supported by the Natural Science Foundation of China (Grant Nos. 61702066 and 11747125), the Chongqing Research Program of Basic Research and Frontier Technology (Grant No. cstc2017jcyjAX0256 and cstc2018jcyjAX0154), and the Research Innovation Program for Postgraduate of Chongqing (Grant Nos. CYS17217 and CYS18238)

Conflicts of Interest: The authors declare no conflict of interest.

References

1. Zhang, Z.J.; Lai, C.F.; Chao, H.C. A green data transmission mechanism for wireless multimedia sensor networks using information fusion. *IEEE Wirel. Commun.* **2014**, *21*, 14–19. [CrossRef]
2. Wu, Y.; Guan, Y.; He, S.; Xin, M. An Industrial-Based Framework for Distributed Control of Heterogeneous Network Systems. *IEEE Trans. Cybern. Syst.* **2018**, *99*, 1–9. [CrossRef]
3. Wu, X.; Wang, H.; Liu, C.; Jia, Y. Cross-View Action Recognition Over Heterogeneous Feature Spaces. *IEEE Trans. Image Process.* **2015**, *24*, 4096–4108. [PubMed]
4. Shan, Z.; Xia, Y.; Hou, P.; He, J. Fusing Incomplete Multisensor Heterogeneous Data to Estimate Urban Traffic. *IEEE MultiMed.* **2016**, *23*, 56–63. [CrossRef]
5. Zhang, Z.; Zou, Y.; Gan, C. Textual sentiment analysis via three different attention convolutional neural networks and cross-modality consistent regression. *Neurocomputing* **2018**, *275*, 1407–1415. [CrossRef]
6. Mantri, D.S.; Prasad, N.R.; Prasad, R. Mobility and Heterogeneity Aware Cluster-Based Data Aggregation for Wireless Sensor Network. *Wirel. Pers. Commun.* **2016**, *86*, 975–993. [CrossRef]
7. Akbar, A.; Khan, A.; Carrez, F.; Moessner, K. Predictive Analytics for Complex IoT Data Streams. *IEEE Internet Things J.* **2017**, *4*, 1571–1582. [CrossRef]

8. Yim, H.J.; Seo, D.; Jung, H.; Back, M.K.; Kim, I.; Lee, K.C. Description and classification for facilitating interoperability of heterogeneous data/events/services in the Internet of Things. *Neurocomputing* **2017**, *256*, 13–22. [CrossRef]
9. Qiu, T.; Chen, N.; Li, K.; Atiquzzaman, M.; Zhao, W. How Can Heterogeneous Internet of Things Build Our Future: A Survey. *IEEE Commun. Surv. Tutor.* **2018**, *20*, 2011–2027. [CrossRef]
10. Yang, J.; Han, Y.; Wang, Y.; Jiang, B.; Lv, Z.; Song, H. Optimization of real-time traffic network assignment based on IoT data using DBN and clustering model in smart city. *Future Gener. Comput. Syst.* **2017**. [CrossRef]
11. Zhang, Q.; Yang, L.T.; Chen, Z.; Xia, F. A High-Order Possibilistic *C*-Means Algorithm for Clustering Incomplete Multimedia Data. *IEEE Syst. J.* **2017**, *11*, 2160–2169. [CrossRef]
12. Han, Y.; Wang, Z.; Li, D.; Guo, Q.; Liu, G. Low-Complexity Iterative Detection Algorithm for Massive Data Communication in IIoT. *IEEE Access* **2018**, *6*, 11166–11172. [CrossRef]
13. Fekade, B.; Maksymyuk, T.; Kyryk, M.; Jo, M. Probabilistic Recovery of Incomplete Sensed Data in IoT. *IEEE Internet Things J.* **2018**, *5*, 2282–2292. [CrossRef]
14. Mendes, L.D.P.; Rodrigues, J.J.P.C.; Lloret, J.; Sendra, S. Cross-Layer Dynamic Admission Control for Cloud-Based Multimedia Sensor Networks. *IEEE Syst. J.* **2014**, *8*, 235–246. [CrossRef]
15. Zhao, L.; Chen, Z.; Yang, Z.; Hu, Y.; Obaidat, M.S. Local Similarity Imputation Based on Fast Clustering for Incomplete Data in Cyber-Physical Systems. *IEEE Syst. J.* **2018**, *12*, 1610–1620. [CrossRef]
16. Zhang, Z.; Zeng, T.; Yu, X.; Sun, S. Social-aware D2D Pairing for Cooperative Video Transmission Using Matching Theory. *Mob. Netw. Appl.* **2018**, *23*, 639–649. [CrossRef]
17. Li, T.; Zhang, L.; Lu, W.; Hou, H.; Liu, X.; Pedrycz, W.; Zhong, C. Interval kernel Fuzzy *C*-Means clustering of incomplete data. *Neurocomputing* **2017**, *237*, 316–331. [CrossRef]
18. Zhang, S.; Yang, Z.; Xing, X.; Gao, Y.; Xie, D.; Wong, H.S. Generalized Pair-Counting Similarity Measures for Clustering and Cluster Ensembles. *IEEE Access* **2017**, *5*, 16904–16918. [CrossRef]
19. Hoecker, M.; Polsterer, K.L.; Kugler, S.D.; Heuveline, V. Clustering of Complex Data-Sets Using Fractal Similarity Measures and Uncertainties. In Proceedings of the 2015 IEEE 18th International Conference on Computational Science and Engineering, Porto, Portugal, 21–23 October 2015; pp. 82–91.
20. Zhou, L.; Wu, D.; Zheng, B.; Guizani, M. Joint physical-application layer security for wireless multimedia delivery. *IEEE Commun. Mag.* **2014**, *52*, 66–72. [CrossRef]
21. Li, F.; Qiao, H.; Zhang, B. Discriminatively boosted image clustering with fully convolutional auto-encoders. *Pattern Recognit.* **2018**, *83*, 161–173. [CrossRef]
22. Gebru, I.D.; Alameda-Pineda, X.; Forbes, F.; Horaud, R. EM Algorithms for Weighted-Data Clustering with Application to Audio-Visual Scene Analysis. *IEEE Trans. Pattern Anal. Mach. Intell.* **2016**, *38*, 2402–2415. [CrossRef]
23. Abualigah, L.M.; Khader, A.T.; Al-Betar, M.A. Unsupervised feature selection technique based on genetic algorithm for improving the Text Clustering. In Proceedings of the 2016 7th International Conference on Computer Science and Information Technology (CSIT), Amman, Jordan, 13–14 July 2016; pp. 1–6.
24. Saadaoui, F.; Bertrand, P.R.; Boudet, G.; Rouffiac, K.; Dutheil, F.; Chamoux, A. A Dimensionally Reduced Clustering Methodology for Heterogeneous Occupational Medicine Data Mining. *IEEE Trans. Nanobiosci.* **2015**, *14*, 707–715. [CrossRef] [PubMed]
25. Zhou, Q. Research on heterogeneous data integration model of group enterprise based on cluster computing. *Clust. Comput.* **2016**, *19*, 1275–1282. [CrossRef]
26. Ramachandran, N.; Perumal, V. Delay-aware heterogeneous cluster-based data acquisition in Internet of Things. *Comput. Electr. Eng.* **2018**, *65*, 44–58. [CrossRef]
27. Meng, L.; Tan, A.H.; Xu, D. Semi-Supervised Heterogeneous Fusion for Multimedia Data Co-Clustering. *IEEE Trans. Knowl. Data Eng.* **2014**, *26*, 2293–2306. [CrossRef]
28. Li, P.; Chen, Z.; Yang, L.T.; Zhao, L.; Zhang, Q. A privacy-preserving high-order neuro-fuzzy *C*-means algorithm with cloud computing. *Neurocomputing* **2017**, *256*, 82–89. [CrossRef]
29. Zhang, Q.; Yang, L.T.; Chen, Z.; Li, P. High-order possibilistic c-means algorithms based on tensor decompositions for big data in IoT. *Inf. Fusion* **2018**, *39*, 72–80. [CrossRef]
30. Mohammadi, M.; Al-Fuqaha, A.; Sorour, S.; Guizani, M. Deep Learning for IoT Big Data and Streaming Analytics: A Survey. *IEEE Commun. Surv. Tutor.* **2018**, *20*, 2923–2960. [CrossRef]
31. Xie, J.; Girshick, R.; Farhadi, A. Unsupervised Deep Embedding for Clustering Analysis. *arXiv* **2015**, arXiv:1511.06335.

32. Kingma, D.P.; Welling, M. Auto-Encoding Variational Bayes. *arXiv* **2013**, arXiv:1312.6114.
33. Li, X.; Chen, Z.; Poon, L.K.M.; Zhang, N.L. Learning Latent Superstructures in Variational Autoencoders for Deep Multidimensional Clustering. *arXiv* **2018**, arXiv:1803.05206.
34. Hou, X.; Shen, L.; Sun, K.; Qiu, G. Deep Feature Consistent Variational Autoencoder. In Proceedings of the 2017 IEEE Winter Conference on Applications of Computer Vision (WACV), Santa Rosa, CA, USA, 24–31 March 2017; pp. 1133–1141.
35. Lopez-Martin, M.; Carro, B.; Sanchez-Esguevillas, A.; Lloret, J. Conditional Variational Autoencoder for Prediction and Feature Recovery Applied to Intrusion Detection in IoT. *Sensors* **2017**, *17*, 1967. [CrossRef] [PubMed]
36. Celikyilmaz, A.; Trksen, I.B. *Modeling Uncertainty with Fuzzy Logic: With Recent Theory and Applications*; Springer Publishing Company: Berlin/Heidelberg, Germany, 2009.
37. Dovžan, D.; Škrjanc, I. Recursive fuzzy C-means clustering for recursive fuzzy identification of time-varying processes. *ISA Trans.* **2011**, *50*, 159–169. [CrossRef] [PubMed]
38. Jérôme, M.; Rui, A.; Francisco, S. Adaptive fuzzy identification and predictive control for industrial processes. *Expert Syst. Appl.* **2013**, *40*, 6964–6975.
39. Rastegar, S.; Araujo, R.; Mendes, J. Online Identification of Takagi-Sugeno Fuzzy Models Based on Self-Adaptive Hierarchical Particle Swarm Optimization Algorithm. *Appl. Math. Model.* **2017**, *45*, 606–620. [CrossRef]

sensors

MDPI

Article

A Real-World Approach on the Problem of Chart Recognition Using Classification, Detection and Perspective Correction

Tiago Araújo [1,2,*], Paulo Chagas [3], João Alves [2], Carlos Santos [1], Beatriz Sousa Santos [2] and Bianchi Serique Meiguins [1,*]

1 Computer Science Graduate Program (PPGCC), Federal University of Pará (UFPA), Belém 66075-110, Brazil; carlossantos@ufpa.br

2 Institute of Electronics and Informatics Engineering of Aveiro (IEETA), Department of Electronics, Telecommunications e Informatics (DETI), University of Aveiro (UA), 3810-193 Aveiro, Portugal; jbga@ua.pt (J.A); bss@ua.pt (B.S.S.)

3 Computer Science Graduate Program (PGCOMP), Federal University of Bahia (UFBA), Salvador 40210-630, Brazil; paulo.chagas@ufba.br

* Correspondence: tiagoaraujo@ufpa.br (T.A.); bianchi@ufpa.br (B.S.M.)

Received: 8 June 2020; Accepted: 13 July 2020; Published: 5 August 2020

Abstract: Data charts are widely used in our daily lives, being present in regular media, such as newspapers, magazines, web pages, books, and many others. In general, a well-constructed data chart leads to an intuitive understanding of its underlying data. In the same way, when data charts have wrong design choices, a redesign of these representations might be needed. However, in most cases, these charts are shown as a static image, which means that the original data are not usually available. Therefore, automatic methods could be applied to extract the underlying data from the chart images to allow these changes. The task of recognizing charts and extracting data from them is complex, largely due to the variety of chart types and their visual characteristics. Other features in real-world images that can make this task difficult are photo distortions, noise, alignment, etc. Two computer vision techniques that can assist this task and have been little explored in this context are perspective detection and correction. These methods transform a distorted and noisy chart in a clear chart, with its type ready for data extraction or other uses. This paper proposes a classification, detection, and perspective correction process that is suitable for real-world usage, when considering the data used for training a state-of-the-art model for the extraction of a chart in real-world photography. The results showed that, with slight changes, chart recognition methods are now ready for real-world charts, when taking time and accuracy into consideration.

Keywords: chart recognition; deep learning; visualization; classification; detection; perspective correction

1. Introduction

Data charts are widely used in technical, scientific, and financial documents, being present in many other subjects of our daily lives, such as newspapers, magazines, web pages, and books. In general, a well-designed data chart leads to an intuitive understanding of its underlying data. In the same way, wrong design choices on chart generation can lead to misinterpretation or later preclude correct data analysis. For example, wrong chart choice and poor mapping of visual variables can reduce the chart quality due to lack of relevant items, such as labels, names of axes, or subtitles. A redesign of those visual representations might be needed to fix these misconceptions.

With the original chart data available, it is possible to perform the necessary changes for mitigating the problems that are presented earlier. However, in the majority of cases, these charts are displayed as static images, which means that the original data are not usually available. Like so, automatic methods could be applied to perform the chart analysis from these images, aiming to obtain the raw data.

When the input image contains other elements besides the chart image (text labels, for example), the detection of these charts must come as a prior step. This detection aims to locate and extract the data chart only, improving recognition performance. Additionally, in a real-world photograph, there is the factor of perspective to take into account. This factor means that the chart can be misaligned and might need some correction for the extraction step. Figure 1 illustrates an example of this situation. The chart is in the middle of a book page and slightly tilted, which indicates that it needs some perspective correction.

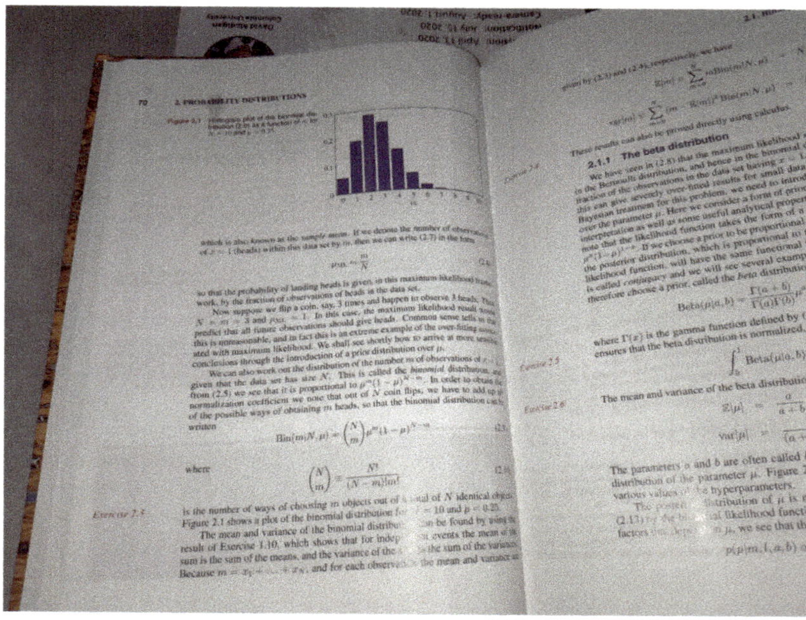

Figure 1. A Bar chart on real-world photography: tilted and in the middle of the text. The image is taken from a book [1]. Modern Chart Recognition methods do not cover real-world situations like this one.

The process of automatic chart data extraction has two main steps [2,3]: chart recognition; and data extraction following a specific extraction algorithm for each type of data chart. This reverse-engineering for chart data extraction from static images is explored in the literature and softwares [2,4–7], with algorithms and methods for many chart types. After the data extraction, it is simple to rebuild the chart, while using a visualization library or software. The main process of automatic data chart extraction in Figure 2 covers the steps from the input image to the interaction with the rebuilt chart. The information visualization pipeline is the key of the main process in the reconstruction of new charts, as it can be directly applied on new environments, being used to overlay a new chart in place of the recognized one or simply store the data. This figure also highlights the fundamental initial step of chart recognition (blue area of Figure 2), as the focus from here onward.

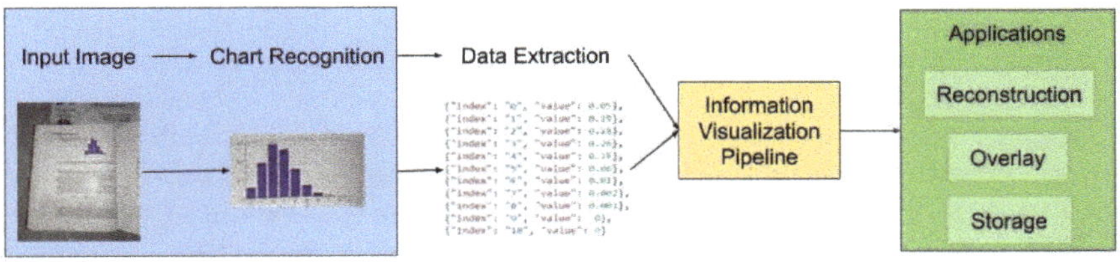

Figure 2. Automatic main process of chart recognition, from the input image to applications. The main process can transform a static environment in a rich user interface for manipulation of data. The information visualization pipeline is environment free, allowing the applications to be used in many environments. This is only possible when the initial step cleans the chart and gives information (blue area).

Research papers regarding chart recognition usually focus on the type classification approach, while assuming a clean image to be classified as a specific chart type. Real-world situations are not that simple, the scenario where a user has a smartphone and wants to manipulate data from the chart is possible, as we have advances in many areas of research that soon will allow for the technology to get in this stage. The features of these scenarios and other real-world usage are not well defined by any modern work of chart recognition.

In this way, chart recognition can be defined as a process that is composed of three computer vision tasks: classification, detection, and perspective correction. Following the literature, the main usage scenario of chart recognition is to discover the chart type [2,3,5,8–11], using it to choose a proper data extraction pipeline. Nevertheless, even without data extraction, there are scenarios that chart recognition can be applied to. Take a set of digital documents as context, like a set of medical papers [12]. In this scenario, it can be useful to create metadata by chart type to support searches.

In the case where the documents have image tags (just as in some PDF and docx files), it is possible to apply classification algorithms to tag these files. However, if the chart is available as a raster-based image only, detection methods should be used. Webpages represent a similar scenario, since an online document can have image tags or SVG-based information for classification. For the printed documents scenario, the detection and perspective correction are fundamental steps to identify and correct chart images. The diagram presented in Figure 3 presents various scenarios and how each step of the process of chart recognition can be used on them.

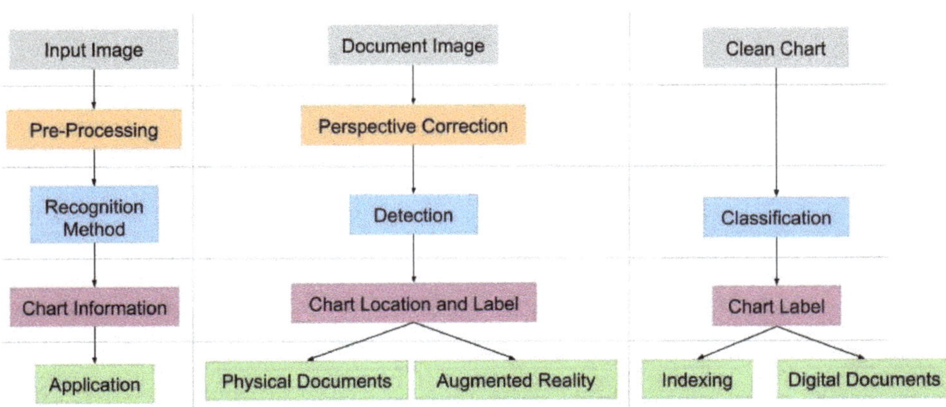

Figure 3. Process for Chart Recognition. Classification is common in literature, but other scenarios can be used if other vision tasks are aggregated. Chart Detection and perspective correction used together can make chart recognition more accurate and usable in new real-world scenarios, like Augmented Reality applications.

Each step of this process of chart recognition brings many difficulties, for example, chart classification is complex due to variations not only between the different types, but also between charts of the same kind, which may differ in data distribution, layout, or presence of noise. Noise removal can be a challenging task, as the environment of the task dictates what is noise. In the context of chart recognition, two scenarios of noise are possible: the noise comes from real-world photography, so lightning and angle could generate an undesired effect on the images, and digitally with resolution loss by image softwares and screens. The other one is noise free, as some charts are generated by visualization libraries or softwares, and are directly integrated in digital documents.

Detection of a chart adds the bounding box search to the classification task, as it needs to locate the charts on a document file, report, or scanned print. Some works have addressed the standard approach for perspective correction through image rectification while using vanishing points [13–16]. However, to the best of our knowledge, none applied to the scenario of chart or document images for chart recognition. Some visual elements may appear in more than one type of chart, hindering the generalization of the classes. For example, consider two recurrent elements of line charts: lines and points. Lines can be found in Area charts, Arc charts, or Parallel Coordinates; points are also present in other chart types, such as scatter plots and some line charts [17]. Additionally, the legends and labels can be mixed with the context of the document, making it hard to locate and correct the chart image.

In the context of chart recognition, we propose an approach for classification, detection, and image rectification on static chart images. This work presents an evaluation of chart classification, paired with experiments on each task of the chart recognition process, namely, detection of charts on document images, and perspective correction using image rectification. The experiments were conducted with state-of-the-art techniques, using datasets of chart images collected from the internet and adapted for each task, with evaluations for each method, in order to analyze the efficiency and efficacy of adding these two steps in the chart recognition process. The results of our experiment presented accuracy in accordance with the most recent challenges in its respective areas. When considering each task's results and the process at hand, it is possible to create applications and models that address the current needs of research on this area, such as preparing chart images for data extraction, image tagging for search, and usage on real-time scenarios. Classification and detection experiments use deep learning models for each step since these methods present outstanding results in several Computer Vision tasks. Furthermore, deep learning methods have been widely used by various chart recognition works [3,5,9,18,19]. Reverse engineering for chart data extraction from static images is explored in the literature and softwares, but it is not the focus of this work. It will be possible to use our process with data extraction methods (use the chart type, position to choose the right algorithms or guide the user, or both), as our process focuses on the starting steps of chart recognition.

Our proposal also used two scenarios of noise removal in the form of perspective correction. Because the classification step is based on clean images with no correction to be done, the detection step uses real document images with clean charts, and the perspective correction step using the distorted chart and document dataset. These scenarios represent the digital documents (no correction) and real-world images (perspective correction).

Moreover, as an example, with the advances of Augmented Reality (AR) technologies [20], it would be possible to recognize charts on the fly by applying the approach proposed here. With a mobile device, users can perform real-time chart recognition on a document and interact with it without changing the context. The usage of AR goggles will allow us to present virtual information directly into the user's field of view while walking on a shopping center, comparing prices, or simply reading a report, providing seamless interactions with charts that were formerly static. This scenario is one use case of edge computing, as some processing could be done on a grander scale on edge than in a traditional cloud service. At the same time, advances are being made on the interaction of AR systems and in nanosystems to allow for in-place processing of these chart recognition models [21].

Despite our work having extensive experiment description sections, the main contributions of our work are to present a whole process for chart recognition that uses many computer vision tasks to cover different tasks. Highlighting the main scenario as recognizing charts in the real world and to present a real-world use case of an example working with no modifications from the trained models.

The organization of this paper is as follows: a rundown on common approaches and terminology for the problem of chart recognition is in Section 2, followed by related works in Section 3. The description of methods is depicted in Section 4, presenting dataset preparation, training regime, and evaluation metrics. The results are in Section 5 and a discussion of chart recognition process based in the results is in Section 6. Final remarks and future works are in Section 7.

2. Chart Recognition

Some Computer Vision tasks are complex, demanding a high level of abstraction and speed, like classification, tracking, and object detection [22]. A natural way to deal with these problems is to use a technique that admits grid-like data as input and does not need a specific feature extractor [23], learning representations with a dataset. Convolutional Neural Network (CNN) is specialized in these requirements and it has achieved excellent results on image classification and other tasks [24].

2.1. Image Classification

In the context of chart image recognition, there has been a focus on the task of image classification, which consists of categorizing a static input image based on a chart image dataset of chart images. Throughout the years, computer vision methods followed the classical methods until the advent of deep learning [2,3].

The image classification task shows the efficiency of representation learning through deep learning as compared to the classical approaches. Classical methods for image classification use handcrafted feature extractors paired with a machine learning classifier. This way, even when the feature extractor is robust, intrinsic spatial data are lost if not explicitly extracted. Furthermore, feature extractors are not universal, so each computer vision problem required manual engineering of features [23]. Thus, CNNs are the current state of the art for image classification tasks.

A CNN groups the filters into hierarchical layers, learning complex and abstract representations from the dataset [25], and orders the filters as layers. State-of-the-art architectures are being used on recent works for chart recognition [3,5,9,19], focusing on the ones that won the ILSVRC challenge [24]. The main ones, which are present in most deep learning textbooks and courses, are: VGG [26], ResNet [27], MobileNet [28] and Inception [29].

The evaluation of these architectures usually follows traditional classification accuracy metric, while using the inference results and comparing with the ground truth labels. In more detail, top 1 accuracy uses the best class of the inference and compares with the ground truth, and the top 5 accuracy uses the range of the best five classes to compare with the target label.

2.2. Object Detection

Object detection is one of the most challenging problems on computer vision, comprehending both classification and localization of objects in an image [30]. In this task, the classification is extended with the bounding boxes' regression, as the identified objects must match in their respectively ground-truth position. A straightforward solution for this task is sliding windows of predefined sizes in all areas of the image and classify each patch. However, this is computationally expensive, making impossible its use on real-world applications. CNN-based solutions can tackle this issue, extending grid-like data processing to object location.

Detection frameworks that are grounded in CNN methods are presenting outstanding results on various object detection domains [24,31,32], as they can learn localization information along with object feature information. These frameworks have a neural network backbone that works as a CNN feature extractor for classification. This backbone can be any CNN model (e.g., Inception, ResNet, or VGG), and its computation can be shared depending on the implementation. There are two main types of these frameworks: one stage detectors and two stage detectors. One stage detectors are fast enough to use in some real-time scenarios, as it speeds up computation without losing accuracy, like RetinaNet [33]. Two-stage detectors usually provide stable results, but they have slower inference time when compared to one-stage detectors, Faster R-CNN [34] being one of them.

The evaluation of object detection frameworks is usually done while using the same metrics of the MS-COCO Recognition Challenge [31], which are the Average Precision (AP) and the Average Recall (AR), both with different scales and thresholds. The AP metric is a relation between precision and recall, and it is separately computed for each class and then averaged. The metric calculation, which uses the true positives and the false positives (for precision and recall), consider two criteria: the predicted class; and the Intersection over Union (IoU) ratio, which measures the overlap of the predicted bounding box and the ground truth bounding box in a certain threshold. If a certain object has its class predicted correctly, and the bounding box predicted IoU is over the threshold, then it is a true positive; otherwise, it is a false positive. The AP is also known as mean Average Precision (mAP), and, in this work, they are equivalent. The equations for AP and IoU are exposed on (1) and (2), respectively, with the variables: C_r the total of classes, c a class, P_b the predicted bounding box, and T_b is the ground truth bounding box.

$$AP = \frac{1}{Cr} \sum_{Cr} \sum_{c} precision(c) \times \Delta recall(c) \tag{1}$$

$$IoU = \frac{Area(P_b \cap T_b)}{Area(P_b \cup T_b)} \tag{2}$$

The COCO challenge defines different average precision notations for different IoU thresholds, notably for an average of 10 values of IoU threshold values ranging from 0.5 to 0.95 with a step of 0.05 (notated as $[0.5 : 0.05 : 0.95]$ from here onward); the 0.50 threshold; and, the 0.75 threshold. These metrics are notated, respectively, as AP (for $IoU = [0.5 : 0.05 : 0.95]$), $AP^{IoU=.5}$ and $AP^{IoU=.75}$.

2.3. Perspective Correction

The perspective correction has been applied in many computer vision tasks, such as automobile license plate recognition, non-Latin characters OCR, and document rectification [35]. These corrections are applied to perspective distortions that can be found in real-world photography, as digital cameras follow a pinhole camera model that generates it [36]. Real-world chart recognition is subject to these perspective distortions and also of its corrections.

Techniques for perspective correction have been widely used in real-world situations in the scenario of planar document rectification, where a distorted document is corrected for future processing, mostly OCR. Chart detection models frameworks could benefit from a rectified document image before performing object detection, as the image comes from a digital camera. For example, it is possible to use some approaches directly of image rectification over photos to chart document rectification.

Image rectification is the reprojection of image planes onto a common plane, and this common plane is parallel to the line between camera centers. Formally, given two images, image rectification determines a transformation of each image plane, such that pairs of conjugate epipolar lines become collinear and parallel to one of the image axes [13]. A way of achieving this is through homography transformations.

The homography is a transformation that defines a relationship of two images on the same plane. This transformation can be used to rectify an image, given relationship hints of the distorted image with its rectified version. One way of achieving this is discovering the vanishing points of an image and using these points to estimate homography between the distorted image and its rectified version.

While vanishing points for homography estimation is present in many methods of image rectification, the method for finding these vanishing points can vary from simple methods to more robust ones. An example of simple methods is matching epipolar lines directly [13] or finding parallel lines [35]. Additionally, robust method examples are searching edgelets [14], using RANSAC on radon transformed images [15] or even training a neural network [37].

Usually, the evaluation is done by comparing images or counting correctly recognized words by OCR software [37–39]. In cases where one has the homography matrices that distorted the images, an evaluation of errors can be measured with an error metric, like Mean Absolute Error (MAE), since it can be used to measure an estimator.

3. Related Works

Several works have been developed on the topic of data chart image classification and detection. These tasks have gained attention, mainly due to its importance in the automatic chart analysis process. Following the traditional image classification pipeline, Savva et al. [2] presents Revision, a system that classifies and extracts data to recreate charts. The dataset used has 2500 images and is collected from the internet and it is composed of 10 classes—area charts, bar charts, curve plots, maps, Pareto charts, pie charts, radar plots, scatter plots, tables, and Venn diagrams. A set of low-level image features and text-level image features were used as input of an SVM classifier, with an average accuracy of 80%. Our work and many others follow this concept of collecting datasets from the internet.

Jung et al. [5] proposed the ChartSense, an interactive system for chart analysis, including the chart classification and data extraction steps. They also used CNNs for classification, comparing three well-known models from the literature: LeNet-1, AlexNet, and GoogLeNet. The models were evaluated while using the Revision dataset. For final classification, more images were collected and added to the Revision dataset, achieving the best accuracy of 91.3% while using GoogLeNet.

Chagas et al. [9] proposed an evaluation of more robust CNN models for chart image classification. Unlike the previously cited works, the proposed methodology has two main tasks: training using synthetically generated images only, comparing the CNN models with conventional classifiers, such as decision trees and support vector machines. The proposed approach aimed to evaluate how the models behave when training with "clean" generated images and testing on noisy internet images. They used a 10-class dataset (arc diagram, area chart, bar chart, line chart, parallel coordinate, pie chart, reorderable Matrix, scatter plot, sunburst, and treemaps) with 12,059 images for training (approximately 1.200 instances for class) and 2683 images from test, evaluating three state-of-art CNN models: VGG-19, Inception-V3, and Resnet-50. The best result was the accuracy of 77.76% while using Resnet-50.

The work of Dai et al. [3] uses few classes (Bar, Pie, Line Scatter, and Radar) than ChartSense, Revision, and the work of Chagas et al., but with accuracy around 99% for all CNNs evaluated. The dataset is also collected from the internet, it has 11,174 images with semi-balanced instances for classes, and the work follows the classification with data extraction. In this context, CNNs showed state-of-the-art results throughout the years for the problem of chart image classification, and our work extends the classes of charts used (10 for 13), followed by a straightforward parameter selection for the state-of-the-art architectures.

Although some works have addressed the chart analysis problem, most of them focused on the chart classification and data extraction tasks, while only a few approached the chart detection issue. Kavasidis et al. [10] introduced a method for automatic multi-class chart detection in document images using a deep-learning approach. Their approach used a trained CNN to detect salient regions of the following object classes: bar charts, line charts, pie charts, and tables. Furthermore, a custom loss function based on the saliency maps was implemented, and a fully-connected Conditional Random Field (CRF) was applied at the end to improve the final predictions. The proposed model was evaluated on the standard ICDAR 2013 dataset (tables only) [40], and on an extended version with new annotations of other chart types. Their best results achieved an average F1-measure of 93.35% and 97.8% on the extended and standard datasets, respectively.

Following a similar path to chart detection, some works have been tackling the table recognition task on document images. Gatos et al. [41] proposed a technique for table detection in document images, including horizontal and vertical line detection. Their approach is only based on image preprocessing and line detection, not requiring any training or heuristics. Schreiber et al. [42] developed the DeepDeSRT, a system for detecting and understanding tables in document images. Their work used Faster R-CNN architecture, which is a state-of-art CNN model for object detection. The proposed model was evaluated on the ICDAR 2013 table competition dataset [40] and a dataset containing documents from a major European aviation company. Document images are used in our work to build a chart detection dataset with chart overlay.

The primary goal of chart detection is finding the localization of the chart image on the input image, which is usually a document page. Huang and Tan [43] proposed a method for locating charts from scanned document pages. The strategy of their work is finding figure blocks from an input image and then to classify this figure in a chart or not. The figure localization used an analysis of logical layout and bounding box, and the image classification is based on statistical features from charts and non-chart elements. Even though their method does not return a specific chart type, the proposed approach achieved promising results, obtaining 90.5% of accuracy on figure location. For figure classification, the results were 91.7% and 94.9% of precision for chart and non-chart classification, respectively. Their work focuses on finding charts and does not fall on the direct definition of multi-class object detection used in our work.

Multi-class chart detection in document images is still an active field of research. One major challenge in this field is defining relevant features for classifying different chart classes, which may vary depending on specialist skills or chart types. This way, deep-learning methods have the advantage of not relying on hand-crafted features or domain-based approaches [23]. Accommodatingly, recent papers have used deep-learning-based architectures for chart classification; in this way, the work presented in this article uses more classes (13) and more images (approx. 21.000) as well as chart detection and perspective correction.

These papers cover specific steps of chart recognition, while allied with some other steps from the main process of chart recognition, extraction, and reconstruction. We take influence on many aspects of these works, like the dataset collection, the classes division, and the chart overlay on document approach, despite that, different from the previous works, our work covers all of the steps from the chart recognition, filling a gap of a complete process to compute static chart image into information. In addition, it also introduces a real-world example of chart recognition of charts on a book.

4. Methods

Most of the choices for the methods used in this work are based on the challenges that emerged from the following tasks, ImageNet for classification, MS-COCO for detection, and ICDAR dewarping for perspective correction. These are hard challenges that proved the efficacy of these models. The methods that are used to train the models, hyperparameter selection, dataset collection, and evaluation are described in the next subsections.

4.1. Datasets

A chart dataset must cover significant differences of each chart type. Data aggregation, background, annotations, and visual marks placement are visual components that vary from chart to chart, even as the same class. This variability is expected and some authors [2,3,5,18,44] address this variability on the collection step, searching the images from the internet, where chart designers publish their work in various different styles. While some datasets could be used for training and evaluation of these techniques, as the ReVision dataset [2] or the MASSVIS dataset [45], we choose to collect data from the internet to use a large number of images to train the methods.

The dataset collection step of this work follows the approach of [3], downloading the images from six web indexers: Google, Baidu, Yahoo, Bing, AOL, and Sogou. The chart types used are arc, area, bar, force-directed graph, line, scatterplot matrix, parallel coordinates, pie, reorderable matrix, scatterplot, sunburst, treemap, and wordcloud with the following keywords (and its chinese translations): arc chart, area chart, bar chart, bars chart, force-directed graph, line chart, scatterplot matrix, parallel coordinates, pie chart, donut chart, reorderable matrix, scatterplot, sunburst chart, treemap, wordcloud, and word cloud. More than 150,000 images were collected using these queries, and we kept only the visualization that falls on the following criteria: two-dimensional (2D) visualizations, not hand drawing, and no repetitions. The total of images downloaded that falls in our rules was 21,099, and the summary of the dataset is in Table 1, with its respective train/test split. The split process was automatically done by a script on the image files, and the training split ranges from 85% to 90%, depending on the number of instances of the class. All 13 classes are used for all of the experiments.

Table 1. Dataset summary, with train and test split by each chart type. This dataset is used throughout all steps, with modifications pertinent to each one of them.

Chart Types	Instances		
	Train	**Test**	**Train + Test**
Arc	129	26	155
Area	494	87	581
Bar	3883	761	4644
Force Directed Graph	1137	228	1365
Line	2618	529	3147
Parallel Coordinates	702	168	870
Pie	2415	481	2896
Reorderable Matrix	242	42	284
Scatterplot	1797	228	2025
Scatterplot Matrix	837	158	995
Sunburst	540	65	605
Treemap	626	73	699
Wordcloud	2557	276	2833
Total	17,977	3122	21,099

The selected types cover most usages of visualizations. The bar chart, line chart, scatterplot, pie chart, and word cloud are chosen, as they are broadly used [4]. Sunburst and treemap are hierarchical visualizations, reorderable matrix, and scatterplot matrix are a multi-facet visualization type. Area and parallel coordinates are multi-dimension visualizations, and arc and force-directed graphs are graph-based visualizations. The selection of these types covers most users' needs. Some classes have few images, as they are not as popular.

The classification experiment uses the downloaded images, ratio scaled and padded to (100×100) size, randomly received augmentation on shear, and zoom by a factor of 0.2 and a 0.5 chance of horizontal flipping, and the pixel values are normalized to be in the –1 and 1 range. For the detection dataset, context insertion is used to create a scenario for chart detection close to a real document page. For this step, the generated charts are overlaid over real document images. Some works used similar approaches, showing results that were at par with the classic approaches [46–48].

The charts were uniformly located entirely in the document image. In some documents, scale transformation is used, by 1/2 or 1/4 of the size of charts. The size of the document images is scaled to 1068×800, where the charts have dimensions that vary from 32×32 to 267×200. The document images used in this work are from the Document Visual Question Answering challenge in the context of CVPR 2020 Workshop on Text and Documents in the Deep Learning Era [49], which features document images for high-level tasks.

A distorted images test dataset of the detection experiment is used for the perspective correction experiment. These distortions are applied while using homography matrices generated with a simple method of perturbation, where a factor of 2 moves each corner of the document image and a homography is calculated with this new distorted image. Figure 4 shows samples of the three datasets.

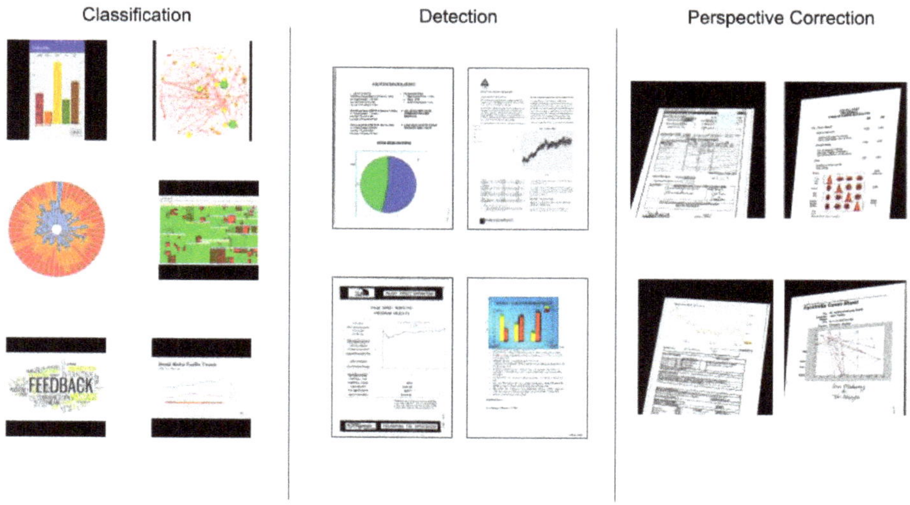

Figure 4. Samples of three datasets for the experiments (from left to right): classification, with added chart images; detection, with chart overlaying document images; and, perspective correction, with distorted images.

4.2. Training and Evaluation

The most common training approach for deep-learning applications uses a pre-trained model and then retrains the model on a new domain dataset. This strategy also applies to CNN classification and detection problems, aiming to exploit features learned on a source domain, leading to a faster and better generalization on a target domain [50]. For this work, the models were pre-trained on the ImageNet dataset [24] for classification and MS COCO [31] for detection. This retraining step is also called fine-tuning, where some (or all) layers must be retrained, adapting the pre-trained model to the chart detection domain. We chose the transfer-learning approach based on fine-tuning the entire network on the target domain. For object detection, this could be done in two ways: with a pre-trained backbone only or with the whole network pre-trained, including the object boxes subnetworks. We chose the pre-trained backbone on ImageNet, because it allows results that reflect some common use cases.

The backbone can be fine-tuned from a large-scale image classification dataset, such as ImageNet. The features can be easily transferred to the new domain, since the backbone is necessarily a set of convolutional layers that can identify features, just like in the classification domain. The subnetworks for box prediction are fine-tuned similarly, but using the knowledge of the region proposal stage (for two-stage detectors) or using the last layers of the convolutional body (for one-stage detectors) to improve box location precision.

For both classification and detection experiments, no mid training changes were used (early stopping, schedule for learning rate changes). They followed the default parameters of the engines unless explicitly stated. The models were trained and evaluated in two different machines, classification and perspective correction in one computer with a GTX 1660 with 6 GB of memory, and the detection experiment ran on a computer with a Titan V video card with 12 GB memory. The engines used for the training (Tensorflow [51] and PyTorch [52]) allow for the training of the models in one machine and run on others with different configurations, given some engine restrictions. It is not decisive for the following sections after training.

4.2.1. Classification

The classification experiment evaluated four different CNN architectures: Xception [53], VGG19 [26], ResNet152 [27], and Mobilenet [28]. These architectures have been chosen, as they are considered to be classic in the literature and they are available in most deep learning frameworks [51,52,54].

Their weights are pre-trained on the ImageNet dataset [24], and Hyperparameter selection is used, the training of the five models for architecture is done in a random search fashion, tunning learning rate, and weight decay with values $[10^{-4}, 10^{-5}, 10^{-6}]$ and $[10^{-6}, 10^{-7}]$, respectively, for 30 epochs in batches of 32 images each.

Classification evaluation is done by measuring accuracy on the test set, picking the best prediction of the CNN. The evaluation is done over all classes, and separately on four classes: bar, pie, line, and scatter. These chart types are popular, and they can be used as an estimate to comparison with other works [2,3,18]. All of the models are evaluated using top-1 accuracy.

Tensorflow 2 [51] is used as the Deep Learning engine for training and evaluation. Datasets are loaded and augmented while using native Tensorflow 2 generators. This experiment ran on a GTX 1660 6 GB video card on an 8 GB memory computer.

4.2.2. Detection

The detection experiment evaluated two distinguished object detectors: RetinaNet [33] and Faster R-CNN [55]. The backbone CNNs are ResNets pre-trained on the ImageNet dataset, and the weights of the whole models were pre-trained on the COCO dataset [31], following the work of the original authors. We choose two one-stage detectors that present state-of-the-art results on COCO and Pascal VOC datasets [56], following our premise of using fast methods for detection inference, in order to enable real-time applications. Hyperparameters of the two detectors are used, as defined by the original authors, only changing the batch size to four images and the iterations for 90,000 (approximately 20 epochs).

The evaluation of these detectors is done while using the COCO challenge metrics alongside inference time. The inference time is a critical metric for object detection, since real-time applications can use fast detection in various tasks. It can be computed as the time in seconds that the framework process the input image and returns the class and the bounding box of the objects on the image. Hence, the frameworks process the input image returning the class and the bounding box of the objects on the image. For this work, the frameworks are evaluated while using the AP, $AP^{IoU=0.5}$, $AP^{IoU=0.75}$, and the inference time.

We used the original authors' recommended engine for the implementation of the selected detectors. RetinaNet and Faster R-CNN frameworks are implemented in the Detectron 2 [57] platform, its implementation is publicly available, runs on the Python language, and it is powered by the PyTorch deep-learning framework. Detectron2 is maintained by the original authors of RetinaNet and Faster R-CNN. This experiment ran on a Titan V 12 GB video card on a 64 GB memory computer.

4.2.3. Perspective Correction

The method for perspective correction follows an image rectification approach. Only one method is evaluated once the ready to use ones are not available, and they are not easy to implement from scratch. Also, commercial approaches have data sharing and usage restrictions. The chosen method is a slight variation of the work of [14], and it is available online in [58]. This method estimates the vanishing points to compute a homography matrix to rectify the original image.

The evaluation applied MAE to measure the estimated homography between the ground truth and the distorted image. The assessment considered three scenarios: raw homography, no scaling, and no translation. Some real-time scenarios could benefit from controlling the scaling and position at will without it being imbursed on the transformation. The experiment ran on a 32 GB memory Intel core i7 machine.

5. Results

We present the results of each individual step using recent state of the art of the art methods. The discussion is provided at the end of the section.

5.1. Classification

The classification step shows remarkable results in different conditions. The best models present results for accuracy over 95% results corresponding to all classes (13) and only four classes. The results for the four classes are overall slightly better than 13 classes, but it uses only chart types with a great number of samples. Table 2 shows the best two models of each architecture. The best model is an Xception with a learning rate of 10^{-4} and a decay of 10^{-6}. The other architectures have an error margin of no more than 3.5% as compared to the best, showing that the moderns architectures could be used if some other task is needed to do so. This result indicates that finetuning the models with little hyperparametrization can deliver good results in this task.

Table 2. Results of Chart Classification. Highlight to Xception network with best accuracy results. Blue cells indicate the right predictions and orange ones indicate high error rate.

Architecture	Learning Rate	Decay	Accuracy–13 Classes	Accuracy–4 Classes
Xception	10^{-4}	10^{-6}	0.954	0.95
		10^{-7}	0.953	0.95
ResNet152	10^{-5}	10^{-6}	0.948	0.95
	10^{-4}	10^{-7}	0.947	0.946
VGG19	10^{-5}	10^{-7}	0.945	0.953
		10^{-6}	0.944	0.945
MobileNet	10^{-4}	10^{-6}	0.926	0.94
	10^{-5}	10^{-7}	0.922	0.923

The confusion matrix prsened in Table 3 shows the best Xception model performance for each class and the most common errors over the test set. The scatterplot matrix chart had the most errors than any chart class, with errors pointing to force-directed graph and scatterplot. This mismatch shows that some characteristics of the layout organization are being lost. Arc charts have no errors, and no other class missed itself for it. The mistake could be a clue of a distinct chart type with little data.

Table 3. Results of Chart Classification Confusion Matrix of the best model.

	Arc	Area	Bar	Force Directed Graph	Line	Parallel Coordinates	Pie	Reorderable Matrix	Scatterplot	Scatterplot Matrix	Sunburst	Treemap	Wordcloud
Arc	26	0	0	0	0	0	0	0	0	0	0	0	0
Area	0	87	0	0	0	0	0	0	0	0	0	0	0
Bar	0	0	728	0	28	1	0	0	1	2	0	1	0
Force Directed Graph	0	0	0	222	1	1	1	0	0	0	0	1	2
Line	0	2	9	1	511	0	4	0	1	1	0	0	0
Parallel Coordinates	0	0	1	0	0	151	0	0	0	6	0	0	0
Pie	0	0	1	0	1	1	164	0	0	1	0	0	0
Reorderable Matrix	0	1	2	0	0	0	0	477	0	1	0	0	0
Scatterplot	0	0	0	0	0	1	1	0	40	0	0	0	0
Scatterplot Matrix	0	0	2	10	16	10	2	0	1	184	0	0	3
Sunburst	0	0	0	2	0	0	0	10	0	1	50	0	2
Treemap	0	0	3	1	0	0	0	0	1	0	0	66	2
Wordcloud	0	0	1	2	0	0	0	1	0	1	0	0	271

5.2. Detection

RetinaNet presented the best values for all APs, endorsing the use of Focal Loss for precision improvement on detection. Furthermore, being a one-stage detector also brought the best result for inference time. More training time could be necessary to achieve better results, as Faster R-CNN is a two-stage detector. The inference time for both methods is below 0.25 seconds per image. Given the high resolution of the images and the framework used alongside the video card, it is acceptable for some applications. Table 4 shows the overview results of the detection experiment.

Table 4. Results of AP, $AP^{IoU=.5}$, and $AP^{IoU=.75}$ inference values and time. RetinaNet has the best results for any AP value and inference time.

Method	AP	$AP^{IoU=0.5}$	$AP^{IoU=0.75}$	Inference Evaluation Time (s/img)
RetinaNet	81.987	91.127	89.428	0.199285
Faster R-CNN	69.68	79.101	77.428	0.210505

The AP results for each class follow the total AP shown in Table 5, except for the arc chart and wordcloud classes. This discrepancy of the values for Faster R-CNN and RetinaNet does not comply with results from the literature on other challenges, where RetinaNet is faster, but Faster R-CNN has better AP [33] overall. Our work showed that RetinaNet got better results with time and AP. We did not make any hyperparametrization besides batch size and number of epochs and this might produce results that are more in line with the expected from the literature with cautious hyperparameter search. However, this is beyond the focus of this work. It is important to notice that this time is of the evaluation alone, and it is not from the next section results.

Table 5. Average Precision (AP) values for each class in RetinaNet and Faster R-CNN. RetinaNet has the best class AP for all class besides arc and wordcloud.

Class	RetinaNet	Faster R-CNN
Arc	86.513	88.52
Area	78.004	76.447
Bar	87.428	82.334
Force Directed Graph	79.746	45.519
Line	83.494	61.618
Scatterplot Matrix	81.072	70.266
Parallel Coordinates	81.669	61.582
Pie	88.26	83.063
Reorderable Matrix	67.69	61.392
Scatterplot	76.751	66.804
Sunburst	76.84	52.785
Treemap	89.843	73.419
Wordcloud	88.52	88.633

5.3. Perspective Correction

The rectification experiment for perspective correction presents three scenarios: the estimation of the raw homography (no changes on any parameter), homography without scale, and homography without translation. The MAE from the raw homography and homography without scale had a similar average, 33.16.

We highlight the results that were obtained with homography without translation, as the average value of 0.12 achieved by the method showed that document positioning on the new rectified plane generates more errors because removing the translation from the estimation removes most of the errors. It is essential to notice that the position is not decisive for this process, once it is only is a preprocessing step for chart detection, and it can be safely ignored in most cases.

5.4. Discussion

The process of chart recognition can be used in many scenarios, such as indexing, storage of data, and real time overlay of information. While many works [3,5,9] focused on chart classification, only a few addressed the chart detection problem on documents [10,11]. The chart detection in documents can use general approaches of other vision tasks for it context, as we used state-of-the-art models and methods of the MS-COCO challenge, and it can be amplified enough to use techniques of document analysis research field, like the approaches of real-world photography in document images. Even so, the first experiment is a chart classification, once some works did it with fewer classes than others [3], using different methods [9], and not presented parameter selection.

Various works have used a dataset collected on the internet, which is more important than the classification method chosen, as the classification step's difference is minimal for each CNN architecture. For example, Chagas et al. [9] used synthetic datasets for training and internet collected for testing, with ten classes, and using the same architectures. The results showed that there was a difference in the accuracy of training and testing with the internet collected dataset is above 15%. In this context, some studies regarding hyperparameter selection must be performed, but it should not be exhaustive given a reasonable amount of data.

The results of the classification experiment showed that state-of-the-art architectures could perform very well given enough data, even for the problem of chart classification with many classes. The work of [3] already showed this with four classes, and we expanded it to 13 classes, while using more recent CNN architectures. The safety that is given by these methods allows for interchangeably using these architectures for other tasks aside chart classification, which usually uses a CNN classifier. For example, ResNets are backbones on many detection frameworks [57]. The ImageNet trained Inception architecture is used on the base example of DeepDream [59] application. The MobileNet architectures [28] are small and fast. The loss function of SRGAN is based on VGG19 feature maps [60]. One could choose the best architecture and train a chart classifier to bootstrap another task.

Despite that detection did not reflect the results of literature, it showed that with little to none hyperparametrization, it is possible to train a detector that acquires AP good enough in document scenarios. Although it lacks a dataset of real-world charts annotated, the training of a method using a chart overlay can be successfully used in real-world scenarios, as shown in the next subsection.

Perspective correction presented good results, with low MAE for the non-translation scenario. Some image rectification solutions are industry-ready, embedded in some applications [61], and using a perspective correction on the process of chart recognition looks a natural next step on the document analysis scenario. The implementation of this process also guarantees that old pipelines do not break, as data extraction methods require rectified images. Other image corrections could also be applied with no extra tooling.

One application of the process of chart recognition process proposed can be a real-time use of these methods, as stated in the introduction. While using a Titan V video card, it is possible to detect charts in almost real-time, so for most high-end specs cards, it is possible to use this process on these time-intensive applications [62]. Even if it is not acceptable for frame-by-frame real-time use in time-intensive applications [63], some shortcuts can be used, such as frame skipping, resolution scaling, and object tracking to minimize the perceived latency for the users. For an augmented reality mobile application, a high-end video card could be part of a cloud service that does the heavy computation, allowing for the mobile device to position the results correctly.

6. Use Case

We propose a task of detecting real-world charts in documents using the models trained in our work, and of the best of our knowledge, there is not any annotated dataset for this task. We chose a simple evaluating metric: full detection and partial detection of a chart image. The first one detects all of the charts and no text outside of it, and the second one detects only part of the chart, or there is some text outside it. Only the highest score of the full detection is used. We choose the Bishop's book [1] as our physical document and manually searched all of the bar charts with axes (most popular chart for several uses [4]), and took photographs of them. The images for this task are displayed in Figure 5.

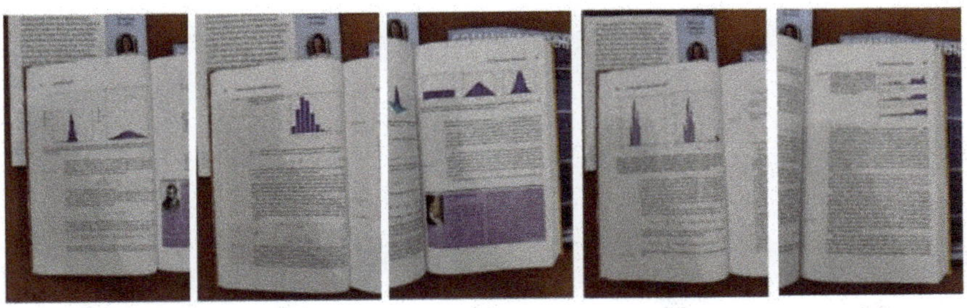

Figure 5. Bar chart photographs taken from a book [1] and transformed for evaluation. (**a**,**d**) present two bar charts with text, (**b**) shows one bar chart, and (**c**,**e**) present three bar charts. For this evaluation, the detector considers only the most accurate detection.

These photographs are transformed by rotations from −4 to 4 degrees with step 0.5, while using the center as the pivot, summing 16 (original + 15 transformations) images for each book page. Two modes are evaluated: a normal mode, with no rectification, and one with rectification, with a total of 160 images at the end. The results are shown on Table 6.

Table 6. Results for chart recognition applied to the images of the book based on two approaches: normal and rectified, for full and partial detection. Each image has 15 other versions, varying by slight rotations. Charts (a) and (d) got no detection in any mode. Rectified images got better detection results for other cases.

Mode	Image	Full	Partial
Camera	Chart (a)	–	–
	Chart (b)	9/16	–
	Chart (c)	6/16	4/16
	Chart (d)	–	–
	Chart (e)	–	–
Rectified	Chart (a)	–	–
	Chart (b)	12/16	–
	Chart (c)	12/16	1/16
	Chart (d)	–	–
	Chart (e)	–	6/16

Even with rectification, some charts are very hard to detect (Figure 5a,d), which implies that even using synthetic overlayed charts, more transformations should be applied. For example, the pure white pages used do not reflect the reality of white from photographs, that receive heavy light influence, as well as more images resolutions to capture the quality of high-end digital cameras. Even so, the rectification results showed that the image preprocessing leads to better results.

Illustrative Example

A single example of a user scenario can showcase the complete step by step chart recognition process. The goal of this example is, given a real-world photograph with a bar chart, to highlight the bar chart position, following the previous use case. This example computes the perspective correction of a real-world photograph with a chart image and detects the position. All steps of this process are executed in a single machine, with a GTX 1060 video card with 6GB of RAM. It is not a high-end video card, but it compensates for its cost. When considering the real world, it is safe to assume that the process will not always have access to the high-end video card specs all the time. The input image is shown in Figure 6.

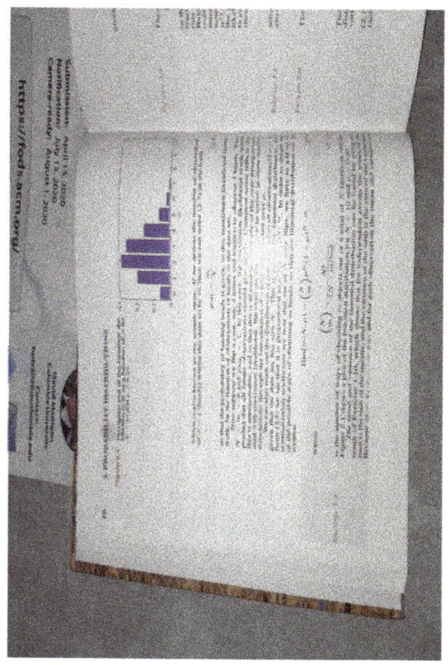

Figure 6. Input image from the use case. The bar chart must be located and extracted.

The first step in this scenario is the perspective correction of the image, so the image rectification method is used. After rectification of the image, the second step is to use the chart detector to recover the chart position and isolate it. These two steps are shown in Figure 7, with its located bar chart.

In total, these two steps took $detection + correction = 0.25 + 0.62 = 0.87$ sec to compute, less time than some camera apps take to save a photography on mobile devices. It is slow to real time frame by frame computation. However, expanding this example, it is possible to use Augmented Reality techniques to superimpose these annotations on the input image directly from the camera stream. Saving the position and using key points of the region makes it possible to track the chart location much faster. In the end, with an extraction method, it is possible to extract the data and highlight it on the image.

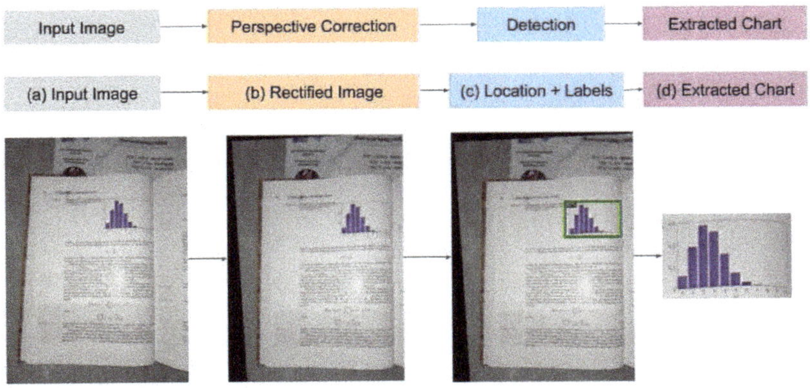

Figure 7. Illustrative example: (**a**) tilted input image, following the (**b**) perspective correction, (**c**) that eased the chart detection (**d**) resulting in a clean-cut bar chart.

Adjustments can be made on the detection model training to recognize charts more accurately, such as introducing different noise options on training, and more training time. However, the results of this use case show that, even with some modifications of state-of-the-art trained models, it is possible to achieve real-time usage of these models. Some hints can be given to the users to position the camera to help the detector. The detection worked without any correction in the simpler cases, but it failed to detect the most tilted charts, and when the detector used the perspective corrected image, it showed a jump in the accuracy of results. In the cases where it is easier to detect, the time of correction could be sparred, but in other hand there is a solution more robust to noise. Our intent with this work is not to show how well the models are trained, but that it is possible to use them on a real-world application of chart recognition chaining these methods.

7. Final Remarks and Future Works

The analysis process of a data chart usually has two main steps: to classify the image into a chart type and to extract data from it. These steps already present several solutions, despite the constant need for better approaches to these tasks. Nevertheless, the majority of these solutions only focus on the classification step, and we have noticed that there is a lack of works in the literature linking real-world photos with the task of labeling charts since before labeling. There are many issues to solve, such as locating charts in images and removing camera distortions. This work presented a modern approach in the process of chart recognition, covering classification, detection, and perspective correction, presenting training methods, dataset collection, and methods already in use by the industry for image rectification. It is the first of this kind, bridging the gap of real-world photography and literature research on the field.

A step little-explored in the literature is detecting the data chart in the image. This step is essential if other elements, such as text or pictures, are present in the image that contains the chart, which is quite common in books, newspapers, and magazines. Along with detection, image rectification could be applied to correct the perspective of documents that contain charts. The experiments presented that, for some scenarios, chart recognition already has the technical toolbox available, but it was not organized on an established process. This work hopes to cover this gap, showing that classification, detection, and perspective correction are ready to be used for initial steps of chart recognition, searching for accuracy or time.

The results of the experiments showed that they individually are pairwise with state of the art chart recognition methods, which is important to validate the main contribution of our work. The perspective correction improved by a significant margin (19 detections of 64 without corrective perspective and 31 of 64 using it) the problem of chart detection for a real-world application. Implying that document noise removal approaches can aid the process of chart recognition.

Future works include adding more visualization types for classification, data extraction algorithms on the process, alongside more image corrections. Lightning and noise are aspects unexplored on this work but they have a wide array of solutions on the document analysis field. The evaluation of more perspective correction methods and how to use them have also be considered. A real-world annotated dataset could help with the assessment of more sophisticated methods, as we proposed in the final sections but lacked the data to make it more robust.

The next generation of mobile devices, paired with high bandwidth of 5G, can launch chart recognition in the real world. This novel process of chart recognition covers the literature and expands it to fill some gaps in real-world applications. For instance, it is possible to create augmented reality applications with a process for chart recognition to be used on new scenarios, creating new research opportunities and challenges.

Author Contributions: Conceptualization, T.A. and B.S.M.; formal analysis, T.A. and P.C.; investigation, C.S.; B.S.M. and B.S.S.; methodology, P.C. and J.A.; project administration, B.S.M.; software, T.A.; supervision, B.S.S. and B.S.M.; validation, P.C.and J.A.; visualization, T.A.; C.S. and B.S.M.; writing—original draft, T.A. and B.S.M.; writing—review & editing, P.C.; J.A.; C.S.; B.S.S. and B.S.M. All authors have read and agreed to the published version of the manuscript.

Funding: This study was financed in part by the Coordenação de Aperfeiçoamento de Pessoal de Nível Superior—Brazil (CAPES)—Finance Code 001 and the APC was funded by the Universidade Federal do Pará (UFPA).

Conflicts of Interest: The authors declare no conflict of interest.

References

1. Bishop, C.M. *Pattern Recognition and Machine Learning*; Springer: Berlin/Heidelberg, Germany, 2006.
2. Savva, M.; Kong, N.; Chhajta, A.; Fei-Fei, L.; Agrawala, M.; Heer, J. Revision: Automated classification, analysis and redesign of chart images. In Proceedings of the 24th annual ACM Symposium on User Interface Software and Technology, Santa Barbara, CA, USA, 16–19 October 2011; pp. 393–402.
3. Dai, W.; Wang, M.; Niu, Z.; Zhang, J. Chart decoder: Generating textual and numeric information from chart images automatically. *J. Vis. Lang. Comput.* **2018**, *48*, 101–109. [CrossRef]
4. Battle, L.; Duan, P.; Miranda, Z.; Mukusheva, D.; Chang, R.; Stonebraker, M. Beagle: Automated extraction and interpretation of visualizations from the web. In Proceedings of the 2018 CHI Conference on Human Factors in Computing Systems, Montreal, QC, Canada, 21–26 April 2018; pp. 1–8.
5. Jung, D.; Kim, W.; Song, H.; Hwang, J.i.; Lee, B.; Kim, B.; Seo, J. ChartSense: Interactive data extraction from chart images. In Proceedings of the 2017 chi Conference on Human Factors in Computing Systems, Denver, CO, USA, May 2017; pp. 6706–6717.
6. Tummers, B. Datathief iii. 2006. Available online: https://datathief.org/ (accessed on 14 July 2020).
7. Mishchenko, A.; Vassilieva, N. Chart image understanding and numerical data extraction. In Proceedings of the 2011 Sixth International Conference on Digital Information Management. IEEE, Melbourn, QLD, Australia, 26–28 September 2011; pp. 115–120.
8. Al-Zaidy, R.A.; Choudhury, S.R.; Giles, C.L. Automatic summary generation for scientific data charts. In Proceedings of the Workshops at the thirtieth aaai Conference on Artificial Intelligence, Phoenix, AZ, USA, 12–13 February 2016.
9. Chagas, P.; Akiyama, R.; Meiguins, A.; Santos, C.; Saraiva, F.; Meiguins, B.; Morais, J. Evaluation of convolutional neural network architectures for chart image classification. In Proceedings of the 2018 International Joint Conference on Neural Networks (IJCNN), Rio de Janeiro, Brazil, 8–13 July 2018; pp. 1–8.
10. Kavasidis, I.; Pino, C.; Palazzo, S.; Rundo, F.; Giordano, D.; Messina, P.; Spampinato, C. A saliency-based convolutional neural network for table and chart detection in digitized documents. In Proceedings of the International Conference on Image Analysis and Processing, Trento, Italy, 9–13 September 2019; pp. 292–302.
11. Svendsen, J.P. Chart Detection and Recognition in Graphics Intensive Business Documents. Ph.D. Thesis, University of Victoria, Victoria, BC, USA, 2015.

12. He, Y.; Yu, X.; Gan, Y.; Zhu, T.; Xiong, S.; Peng, J.; Hu, L.; Xu, G.; Yuan, X. Bar charts detection and analysis in biomedical literature of PubMed Central. In Proceedings of the AMIA Annual Symposium Proceedings. American Medical Informatics Association, Washington, DC, USA, 4–8 November 2017; Volume 2017, p. 859.
13. Fusiello, A.; Trucco, E.; Verri, A. A compact algorithm for rectification of stereo pairs. *Mach. Vis. Appl.* **2000**, *12*, 16–22. [CrossRef]
14. Chaudhury, K.; DiVerdi, S.; Ioffe, S. Auto-rectification of user photos. In Proceedings of the 2014 IEEE International Conference on Image Processing (ICIP), Paris, France, 27–30 October 2014; pp. 3479–3483.
15. Takezawa, Y.; Hasegawa, M.; Tabbone, S. Robust perspective rectification of camera-captured document images. In Proceedings of the 2017 14th IAPR International Conference on Document Analysis and Recognition (ICDAR), Kyoto, Japan, 9–15 November 2017, Volume 6, pp. 27–32.
16. Shemiakina, J.; Konovalenko, I.; Tropin, D.; Faradjev, I. Fast projective image rectification for planar objects with Manhattan structure. In Proceedings of the Twelfth International Conference on Machine Vision (ICMV 2019), Amsterdam, The Netherlands, 6–18 November 2019; p.114331N.
17. Khan, M.; Khan, S.S. Data and information visualization methods, and interactive mechanisms: A survey. *Int. J. Comput. Appl.* **2011**, *34*, 1–14.
18. Tang, B.; Liu, X.; Lei, J.; Song, M.; Tao, D.; Sun, S.; Dong, F. Deepchart: Combining deep convolutional networks and deep belief networks in chart classification. *Signal Process.* **2016**, *124*, 156–161. [CrossRef]
19. Junior, P.R.S.C.; De Freitas, A.A.; Akiyama, R.D.; Miranda, B.P.; De Araújo, T.D.O.; Dos Santos, C.G.R.; Meiguins, B.S.; De Morais, J.M. Architecture proposal for data extraction of chart images using Convolutional Neural Network. In Proceedings of the 2017 21st International Conference Information Visualisation (IV), London, UK, 11–14 July 2017; pp. 318–323.
20. Linowes, J.; Babilinski, K. *Augmented Reality for Developers: Build Practical Augmented Reality Applications with Unity, ARCore, ARKit, and Vuforia*; Packt Publishing Ltd.: Birmingham, UK, 2017.
21. Passian, A.; Imam, N. Nanosystems, Edge Computing, and the Next Generation Computing Systems. *Sensors* **2019**, *19*, 4048. [CrossRef] [PubMed]
22. Parker, J.R. *Algorithms for Image Processing and Computer Vision*; John Wiley & Sons: Hoboken, NJ, USA, 2010.
23. LeCun, Y.; Bengio, Y.; Hinton, G. Deep learning. *Nature* **2015**, *521*, 436–444. [CrossRef] [PubMed]
24. Russakovsky, O.; Deng, J.; Su, H.; Krause, J.; Satheesh, S.; Ma, S.; Huang, Z.; Karpathy, A.; Khosla, A.; Bernstein, M.; others. Imagenet large scale visual recognition challenge. *Int. J. Comput. Vis.* **2015**, *115*, 211–252. [CrossRef]
25. Krizhevsky, A.; Sutskever, I.; Hinton, G.E. Imagenet classification with deep convolutional neural networks. In Proceedings of the Advances in Neural Information Processing Systems, Lake Tahoe, NV, USA, 3–6 December 2012; pp. 1097–1105.
26. Simonyan, K.; Zisserman, A. Very Deep Convolutional Networks for Large-Scale Image Recognition. In Proceedings of the International Conference on Learning Representations, San Diego, CA, USA, 7–9 May 2015.
27. He, K.; Zhang, X.; Ren, S.; Sun, J. Identity mappings in deep residual networks. In Proceedings of the European Conference on Computer Vision, Amsterdam, The Netherlands, 11–14 October 2016; pp. 630–645.
28. Sandler, M.; Howard, A.; Zhu, M.; Zhmoginov, A.; Chen, L.C. Mobilenetv2: Inverted residuals and linear bottlenecks. In Proceedings of the IEEE Conference on Computer Vision and Pattern Recognition, Salt Lake City, UT, USA, 18-22 June 2018; pp. 4510–4520.
29. Szegedy, C.; Ioffe, S.; Vanhoucke, V.; Alemi, A.A. Inception-v4, inception-resnet and the impact of residual connections on learning. In Proceedings of the Thirty-first AAAI conference on artificial intelligence, Francisco, CA, USA, 4–9 February 2017.
30. Girshick, R.; Donahue, J.; Darrell, T.; Malik, J. Region-based convolutional networks for accurate object detection and segmentation. *IEEE Trans. Pattern Anal. Mach. Intell.* **2015**, *38*, 142–158. [CrossRef] [PubMed]
31. Vinyals, O.; Toshev, A.; Bengio, S.; Erhan, D. Show and tell: Lessons learned from the 2015 mscoco image captioning challenge. *IEEE Trans. Pattern Anal. Mach. Intell.* **2016**, *39*, 652–663. [CrossRef] [PubMed]
32. Everingham, M.; Van Gool, L.; Williams, C.K.; Winn, J.; Zisserman, A. The pascal visual object classes (voc) challenge. *Int. J. Comput. Vis.* **2010**, *88*, 303–338. [CrossRef]
33. Lin, T.Y.; Goyal, P.; Girshick, R.; He, K.; Dollár, P. Focal loss for dense object detection. In Proceedings of the IEEE international Conference on Computer Vision, Venice, Italy, 22–29 October 2017; pp. 2980–2988.

34. Girshick, R. Fast r-cnn. In Proceedings of the IEEE international Conference on Computer Vision, Santiago, Chile, 11–18 December 2015; pp. 1440–1448.
35. Jagannathan, L.; Jawahar, C. Perspective correction methods for camera based document analysis. In Proceedings of the First Int. Workshop on Camera-based Document Analysis and Recognition, Seoul, Korea, 29 August–1 September 2005; pp. 148–154.
36. Li, X.; Zhi, Y.; Yin, P.; Duan, C. Camera model and parameter calibration. *E&ES* **2020**, *440*, 042099.
37. Sheshkus, A.; Ingacheva, A.; Arlazarov, V.; Nikolaev, D. HoughNet: neural network architecture for vanishing points detection. *arXiv* **2019**, arXiv:1909.03812.
38. Arlazarov, V.V.; Bulatov, K.B.; Chernov, T.S.; Arlazarov, V.L. MIDV-500: a dataset for identity document analysis and recognition on mobile devices in video stream. *arXiv* **2019**, arXiv:1807.05786.
39. El Abed, H.; Wenyin, L.; Margner, V. International conference on document analysis and recognition (ICDAR 2011)-competitions overview. In Proceedings of the 2011 International Conference on Document Analysis and Recognition, Beijing, China, 18–21 September 2011; pp. 1437–1443.
40. Göbel, M.; Hassan, T.; Oro, E.; Orsi, G. ICDAR 2013 table competition. In Proceedings of the 2013 12th International Conference on Document Analysis and Recognition, Washington, DC, USA, 25–28 August 2013; pp. 1449–1453.
41. Gatos, B.; Danatsas, D.; Pratikakis, I.; Perantonis, S.J. Automatic table detection in document images. In Proceedings of the International Conference on Pattern Recognition and Image Analysis, Genoa, Italy, 7–11 September 2005; pp.609–618.
42. Schreiber, S.; Agne, S.; Wolf, I.; Dengel, A.; Ahmed, S. Deepdesrt: Deep learning for detection and structure recognition of tables in document images. In Proceedings of the 2017 14th IAPR International Conference on Document Analysis and Recognition (ICDAR), Kyoto, Japan, 9–15 November 2017, Volume 1; pp. 1162–1167.
43. Huang, W.; Tan, C.L. Locating charts from scanned document pages. In Proceedings of the Ninth International Conference on Document Analysis and Recognition (ICDAR 2007), Parana, Brazil, 23–26 September 2007; Volume 1, pp. 307–311.
44. Poco, J.; Heer, J. Reverse-engineering visualizations: Recovering visual encodings from chart images. In *Computer Graphics Forum*; Wiley Online Library: Hoboken, NY, USA, 2017; Volume 36, pp. 353–363.
45. Bylinskii, Z.; Borkin, M. Eye fixation metrics for large scale analysis of information visualizations. ETVIS Work. *Eye Track. Vis.* **2015**.
46. Barth, R.; IJsselmuiden, J.; Hemming, J.; Van Henten, E. Synthetic bootstrapping of convolutional neural networks for semantic plant part segmentation. *Comput. Electron. Agric.* **2019**, *161*, 291–304. [CrossRef]
47. Shatnawi, M.; Abdallah, S. Improving handwritten arabic character recognition by modeling human handwriting distortions. *ACM Trans. Asian Low-Resource Lang. Inf. Proc.* **2015**, *15*, 1–12. [CrossRef]
48. Eggert, C.; Winschel, A.; Lienhart, R. On the benefit of synthetic data for company logo detection. In Proceedings of the 23rd ACM international conference on Multimedia, Mountain View, CA, USA, 23–27 October 2015; pp. 1283–1286.
49. CVPR2020 Workshop on Text and Documents in the Deep Learning Era. Available online: https://cvpr2020text.wordpress.com/ (accessed on 8 April 2020).
50. Goodfellow, I.; Bengio, Y.; Courville, A. *Deep Learning*; MIT Press: Cambridge, MA, USA, 2016.
51. Abadi, M.; Agarwal, A.; Barham, P.; Brevdo, E.; Chen, Z.; Citro, C.; Corrado, G.S.; Davis, A.; Dean, J.; Devin, M.; et al. TensorFlow: Large-Scale Machine Learning on Heterogeneous Systems. *arXiv*, **2016**, arXiv:1603.04467.
52. Paszke, A.; Gross, S.; Massa, F.; Lerer, A.; Bradbury, J.; Chanan, G.; Killeen, T.; Lin, Z.; Gimelshein, N.; Antiga, L.; et al. PyTorch: An Imperative Style, High-Performance Deep Learning Library. In *Advances in Neural Information Processing Systems 32*; Wallach, H., Larochelle, H., Beygelzimer, A., d'Alché-Buc, F., Fox, E., Garnett, R., Eds.; Curran Associates, Inc.: Red Hook, NY, USA, 2019; pp. 8024–8035.
53. Chollet, F. Xception: Deep learning with depthwise separable convolutions. In Proceedings of the IEEE Conference on Computer Vision and Pattern Recognition, Honolulu, HI, USA, 21–26 July 2017; pp. 1251–1258.
54. Chen, T.; Li, M.; Li, Y.; Lin, M.; Wang, N.; Wang, M.; Xiao, T.; Xu, B.; Zhang, C.; Zhang, Z. Mxnet: A flexible and efficient machine learning library for heterogeneous distributed systems. *arXiv* **2015**, arXiv:1512.01274.
55. Ren, S.; He, K.; Girshick, R.; Sun, J. Faster r-cnn: Towards real-time object detection with region proposal networks. In Proceedings of the Advances in Neural Information Processing Systems, Montreal, QC, Canada, 7–12 December 2015; pp. 91–99.

56. Liu, L.; Ouyang, W.; Wang, X.; Fieguth, P.; Chen, J.; Liu, X.; Pietikäinen, M. Deep learning for generic object detection: A survey. *Int. J. Comput. Vis.* **2020**, *128*, 261–318. [CrossRef]
57. Wu, Y.; Kirillov, A.; Massa, F.; Lo, W.Y.; Girshick, R. Detectron2. 2019 Available online: https://github.com/facebookresearch/detectron2.
58. Image-Rectification. Available online: https://github.com/chsasank/Image-Rectification (accessed on 8 April 2020).
59. Mordvintsev, A.; Olah, C.; Tyka, M. Deepdream-a code example for visualizing neural networks. *Google Resarch*, 2015, Available online: https://ai.googleblog.com/2015/07/deepdream-code-example-for-visualizing.html (accessed on 8 April 2020) .
60. Ledig, C.; Theis, L.; Huszár, F.; Caballero, J.; Cunningham, A.; Acosta, A.; Aitken, A.; Tejani, A.; Totz, J.; Wang, Z.; others. Photo-realistic single image super-resolution using a generative adversarial network. In Proceedings of the IEEE Conference on Computer Vision and Pattern Recognition, Honolulu, HI, USA, 21–26 July 2017; pp. 4681–4690.
61. Get Office Lens—Microsoft Store. Available online: https://www.microsoft.com/en-us/p/office-lens/9wzdncrfj3t8/ (accessed on 8 April 2020).
62. Feng, X.; Jiang, Y.; Yang, X.; Du, M.; Li, X. Computer vision algorithms and hardware implementations: A survey. *Integration* **2019**, *69*, 309–320. [CrossRef]
63. Raaen, K.; Kjellmo, I. Measuring latency in virtual reality systems. In Proceedings of the International Conference on Entertainment Computing, Tsukuba City, Japan, 18–21 September 2015; pp. 457–462.

MDPI
St. Alban-Anlage 66
4052 Basel
Switzerland
Tel. +41 61 683 77 34
Fax +41 61 302 89 18
www.mdpi.com

Sensors Editorial Office
E-mail: sensors@mdpi.com
www.mdpi.com/journal/sensors

www.ingramcontent.com/pod-product-compliance
Lightning Source LLC
Chambersburg PA
CBHW061629210926
51850CB00009B/2481
9783036530260